Synthesis Lectures on Green Energy and Technology

Series Editor

Zhongchao Tan, Department of Mechanical and Mechatronics Engineering, University of Waterloo, Waterloo, Canada

This series publishes short books in environmental-friendly and sustainable technologies, in particular state-of-the-art research and development in green energy and pollution control. The books in this series are written by top researchers and professionals. The series addresses researchers, advanced students, technical consultants as well as decision makers in industries and politics. Hence, the level of presentation spans from instructional to highly technical.

Yi Zhang

Superhydrophobicity of Surfaces Dressed by Electrospun Fibers

Springer

Yi Zhang
Department of Mechanical and Mechatronics
Engineering
University of Waterloo
Waterloo, ON, Canada

ISSN 2948-2720 ISSN 2948-2739 (electronic)
Synthesis Lectures on Green Energy and Technology
ISBN 978-3-031-55551-0 ISBN 978-3-031-55552-7 (eBook)
https://doi.org/10.1007/978-3-031-55552-7

This Springer imprint is published by the registered company Springer Nature Switzerland AG
The registered company address is: Gewerbestrasse 11, 6330 Cham, Switzerland

Paper in this product is recyclable.

Contents

Abbreviations

CNT	Carbon nanotube
DCM	Dichloromethane
DMAc	*N*, *N*-dimethylacetamide
DMF	N, N-dimethylformamide
DMDCS	Dichlorodimethyl silane
DTMS	*n*-dodecyltrimethoxysilane
FAS	Fluoroalkylsilane
FDTS	(heptadecafluoro-1, 1, 2, 2-tetrahydrodecyl)trichlorosilane
FPI	Fluorinated polyimide
FPU	Fluorinated polyurethane
Hap	Hydroxyapatite
HFIP	Hexafluoro isopropanol
NF	Nanoflake
NP	Nanoparticle
NFES	Near-field electrospinning
PA6	Polyamide 6
PAA	Poly(acrylic acid)
PAH	Poly(allylamine hydrochloride)
PAM	Poly(acryl amide)
PAN	Polyacrylonitrile
PANI	Polyaniline
PCC	n-octadecane phase change capsules
PCL	Poly(caprolactone)
PDMS	Poly(methylhydrosiloxane)
PEO	Poly(ethylene oxide)
PES	Polyethersulfone
PET	Poly(ethylene terephthalate)
PFDT	1H,1H,2H,2H-perfluorodecanethiol
PFDTES	Perfluorodecyltrietho-xysilane
PI	Polyimide

PLLA	Poly(L-lactic acid)
PMMA	Poly(methyl methacrylate)
PnPrOx	Poly(2-n-propyl-2-oxazoline)
PolyPhe	Poly(L-phenylalanine)
PPFEMA	Poly(perfluoroalkyl ethyl methacrylate)
PS	Polystyrene
PS-PDMS	Poly(styrene-*b*-dimethylsiloxane)
PTFE	Poly(tetrafluoroethylene)
PU	Polyurethane
PVA	Poly(vinyl alcohol)
PVB	Polyvinyl butyral
PVC	Polyvinyl chloride
PVDF	Poly(vinylidene fluoride)
PVDF-HFP	Poly(vinylidene fluoride-co-hexafluoropropylene)
PVP	Poly(vinyl pyrrolidone)
SEBS	Poly(styrene-*b*-(ethylene-*co*-butylene)-*b*-styrene)
THF	Tetrahydrofuran
TM	Tourmaline
γ-PGA-Phe	γ-PGA modified with *L*-phenylalanine ethylester

Acronyms

ACA	Advancing contact angle
BN	Boron nitride
CA	Cellulose acetate
CAH	Contact angle hysteresis
CB	Cassie-Baxter
CVD	Chemical vapor deposition
ECA	Equilibrium contact angle
iCVD	Initiated chemical vapor deposition
ICA	Intrinsic contact angle
LBL	Layer-by-layer
LBM	Lattice Boltzmann Method
MCA	Metastable contact angle
RCA	Receding contact angle
RH	Relative humidity
SEM	Scanning electron microscope
WBS	Waterproof, breathable, superhydrophobic
WVTR	Water vapor transmission rate

Symbols

A	Surface area (m^2)
A_1	Surface area of the outer surface of droplet (m^2)
A_2	Apparent contact area between droplet and solid surface (m^2)
A_c	Cup opening area for water vapor transmission rate test (m^2)
d	Fiber diameter (m)
d_c	Characteristic diameter of the produced porous fibers (m)
E	Free energy per unit depth (J/m)
f_1	Ratio of solid-liquid interface area to droplet basal area (–)
f_2	Ratio of bottom liquid-vapor interface area to droplet basal area (–)
f	Ratio of solid-liquid interface area to droplet basal area when $f_1+f_2 = 1$ (–)
f_p	Fraction of fiber surface occupied by pores (–)
G	Free energy (J)
g	Gravity acceleration (m/s^2)
h	Height of surface structure (m)
l	Inter-fiber distance (m)
l_1	Total width of one surface structure and groove (m)
l_2	Width of one surface structure (m)
l_w	Normalized wetting length (–)
L	Length of interface (m)
L_t	Total width from the first to the Nth measured fiber (m)
m	Weight (kg)
N	Number of fibers across the measured area (–)
n	Half number of covered structure unit (–)
r	Radius of surface structure (m)
r_w	Ratio of actual and projected solid-liquid interface area (–)
R_0	Radius of droplet before contact (m)
R_1	Radius of droplet after contact (m)
R_2	Half-length of contact interface (m)
s	Wetted length on horizontal surfaces by exterior TPCL in one structure unit (m)

s'	Wetted length on each horizontal surface by exterior TPCL (m)
S	Cross-section area of droplets (m^2)
$S_{structure}$	Cross-section area of one surface structure (m^2)
S_{groove}	Cross-section area of one groove (m^2)
S_2	Cross-section area difference between the liquid below external TPCL and the structure above external TPCL (m^2)
t	Time (s)
V	Droplet volume (m^3)

Greek Letters

α_r	Roll-off angle (–)
γ	Surface tension (N/m)
δ_i	Wetted length by interior TPCL (m)
δ_e	Wetted length on each inclined surface by exterior TPCL (m)
δ_o	Wetted length on inclined surfaces by exterior TPCL in one structure unit (m)
δ_o'	Wetted length on each inclined surface by exterior TPCL (m)
θ_A	Apparent contact angle (–)
θ_Y	Intrinsic contact angle (–)
$\theta_{Y,F}$	Intrinsic contact angle of fiber (–)
$\theta_{Y,S}$	Intrinsic contact angle of substrate (–)
θ_i	Wetted angle by interior TPCL (–)
θ_o	Angular position of exterior TPCL (–)
θ_{s1}	Structure parameter as defined in Fig. 3.5 (–)
θ_{s2}	$2\varphi-\theta_{s1}$ for $\theta_o<180°$; $-\theta_{s1}$ for $\theta_o>180°$ (–)
θ_{s3}	0 for $\theta_o<180°$; $\varphi-\theta_{s1}$ for $\theta_o>180°$ (–)
ρ	Droplet density (kg/m^3)
φ	Internal angle of surface structure (–)

Subscripts

C1, C2	Position of exterior TPCL in composite state (–)
CB	Cassie-Baxter state (–)
i, ii′, ii″	State (i, ii′, ii″) of droplets on biaxial structure (–)
LV1	Liquid-vapor interface 1 (–)
LV2	Liquid-vapor interface 2 (–)
SL	Solid-liquid interface (–)
SV	Solid-vapor interface (–)

T1-3	Position of interior TPCL in wetting transition (–)
W	Wenzel state (–)
W1-4	Position of exterior TPCL in noncomposite state (–)

Introduction

1

1.1 Background

Liquid wetting of solid surfaces is an ubiquitous phenomenon in nature [1]. The wettability of solid surfaces, which is often assessed by sessile droplet contact angles [2], is an important property to various applications related to energy [3], environment [4], health [5], etc.

Figure 1.1 shows the classification of surface wettability based on the contact angle: (A) superhydrophilicity, (B) hydrophilicity, (C) hydrophobicity, and (D) superhydrophobicity. Specifically, a hydrophilic surface has a strong affinity to liquid, thereby showing a contact angle less than 90°. The surface is superhydrophilic if its contact angle is lower than 5°–10° [6]. Conversely, a hydrophobic surface repels liquid and shows a contact angle greater than 90°. When the contact angle is greater than 150°, the surface can be defined as a "lotus" or "rose-petal" superhydrophobic surface depending on its adhesion strength to droplets. For the lotus superhydrophobicity, as shown in Fig. 1.1(D-1), the sliding angle, which is the threshold tilting angle of the surface when droplets start to slide, is usually less than 10° [7]. In contrast, the rose-petal superhydrophobicity has strong droplet adhesion, so droplets stay pinned on the surface even by turning upside down [8], as shown in Fig. 1.1(D-2).

Figure 1.2 demonstrates that water droplets roll off (A) lotus leaf while rain droplets are retained on (B) rose petals. The lotus superhydrophobicity, also known as "lotus effect", originates from the extreme water repellency and the self-cleaning property of lotus leaf (*Nelumbo nucifera*) [9, 10]. Similarly, the rose-petal superhydrophobicity, or "petal effect", is defined based on the phenomenon that water droplets remain spherical on rose petals but cannot roll off even by turning the petals upside down [11].

Both lotus effect and petal effect have received much attention for their wide applications. For the lotus effect, the ability to remove surface contaminants such as dust particles

Y. Zhang, *Superhydrophobicity of Surfaces Dressed by Electrospun Fibers*, Synthesis Lectures on Green Energy and Technology, https://doi.org/10.1007/978-3-031-55552-7_1

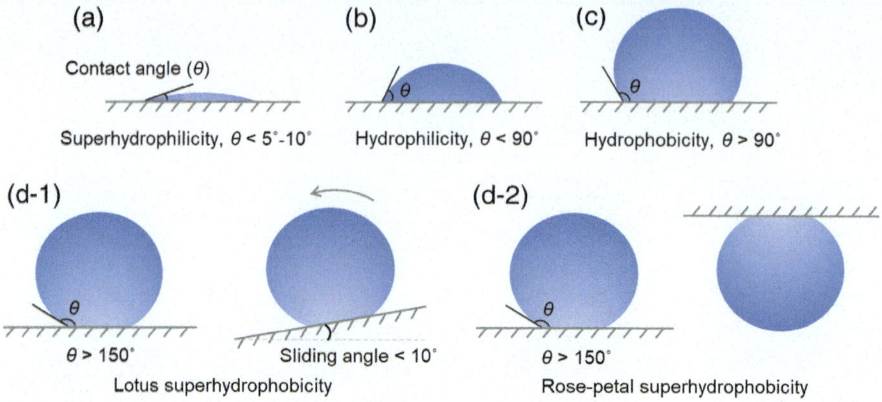

Fig. 1.1 Classification of solid surface wettability based on contact angle: **a** superhydrophilicity, **b** hydrophilicity, **c** hydrophobicity, and **d** superhydrophobicity

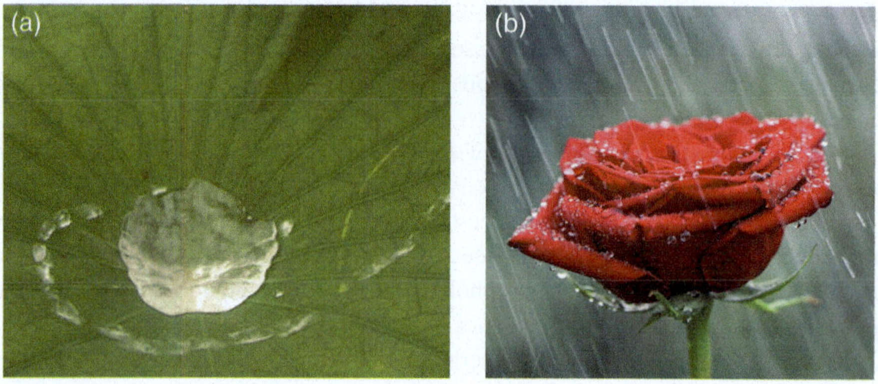

Fig. 1.2 Droplets on **a** lotus leaf [12] and **b** rose petals [13]

using droplets allows for developing self-cleaning surfaces; its applications can be found in any field where clean surfaces are necessary, such as sensors, solar cells, textiles [14]. Moreover, the surfaces with lotus-effect also exhibit properties of anti-corrosion [15], anti-icing [16], anti-fogging [17], etc. For the petal effect, retaining spherical droplets with strong adhesion are promising for microdroplet transportation [18], lap-on-chip devices [19], single molecule spectroscopy [20], etc.

1.2 Wetting Theory

To help understand the physics behind various wetting phenomena, the basic wetting theory for solid surfaces is introduced in this section.

1.2.1 Wetting on Ideal Flat Surface

Figure 1.3 presents the schematic of a droplet wetting on an ideal flat surface. The apparent contact angle (APCA, θ_A), which is the macroscopically observed contact angle [21], can be expressed by the well-known Young's equation [22, 23]:

$$\cos\theta_A = \frac{\gamma_{SV} - \gamma_{SL}}{\gamma_{LV}} \tag{1.1}$$

where γ denotes surface tension; the subscripts S, L, and V represent solid, liquid, and vapor, respectively. The APCA for a flat surface is also known as the intrinsic contact angle (ICA, θ_Y) [24] of the surface material, which is determined by the chemical composition of the surface. Note that surface tension varies among different liquids, so the contact angle for a certain solid surface varies with different probing liquids. In the following part of this book, the contact angle refers to water contact angle unless specified otherwise.

Figure 1.4 shows the schematic of the change of Gibbs free energy with APCA on a flat surface. From a thermodynamic perspective, the spreading of a droplet on a solid surface can be regarded as a Gibbs free energy minimization process with the movement of three-phase contact line (TPCL, see Fig. 1.3). The movement of TPCL along the solid surface changes APCA. The three-phase system is at the global minimum of free energy when θ_Y is reached. Therefore, θ_Y is the equilibrium contact angle (ECA) for flat surfaces.

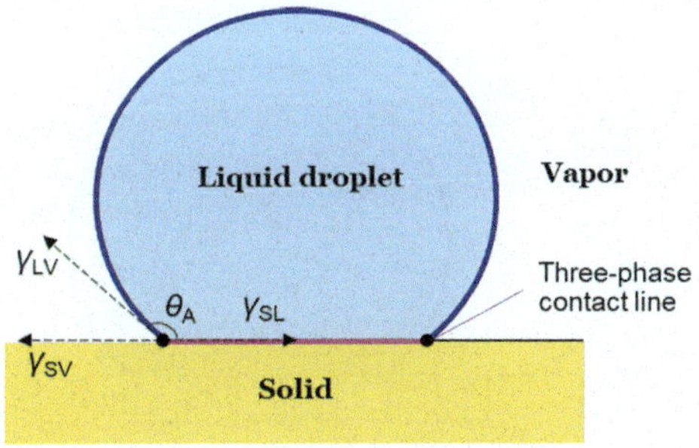

Fig. 1.3 Schematic of droplet wetting on an ideal flat surface

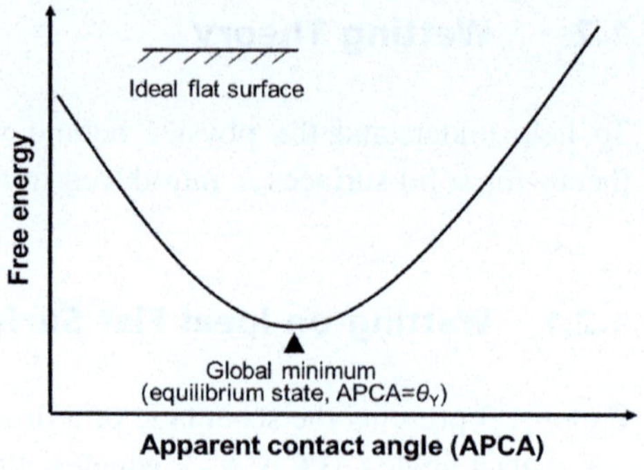

Fig. 1.4 Schematic of the change of free energy with APCA on flat surfaces

1.2.2 Wetting on Rough Surface

In reality, most surfaces exhibit a certain degree of roughness. Figure 1.5 shows two potential wetting states on rough surfaces: (A) composite state and (B) noncomposite state.

In composite state, also known as Cassie-Baxter state [25], air is trapped under the droplet, and the contact angle θ_A can be modeled using Cassie-Baxter equation [25]:

$$\cos\theta_A = f_1 \cos\theta_Y - f_2 \tag{1.2}$$

where f_1 and f_2 are the ratios of solid-liquid and liquid-vapor interface areas, respectively, under the droplet to the droplet basal area. In Fig. 1.5a, f1 $= l_2/l_1$, $f_2 = 1 - l_2/l_1$ given that the structure surface has negligible roughness.

In noncomposite state, or Wenzel [26] state, the solid surface under the droplet is fully wetted by liquid, and the contact angle θ_A can be analyzed using Wenzel equation [26]:

$$\cos\theta_A = r_w \cos\theta_Y \tag{1.3}$$

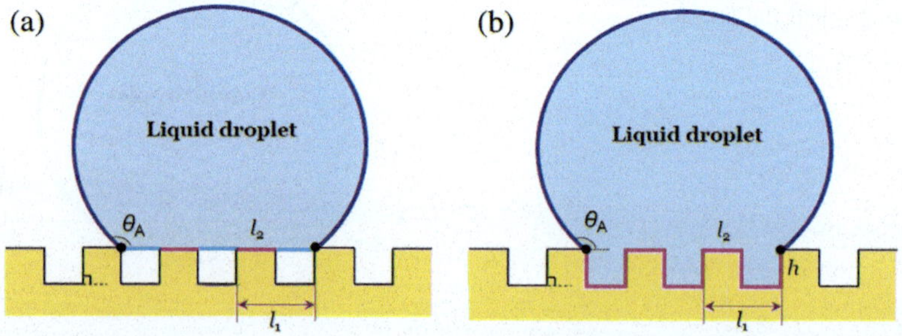

Fig. 1.5 Schematics of **a** composite wetting and **b** noncomposite wetting on rough surfaces

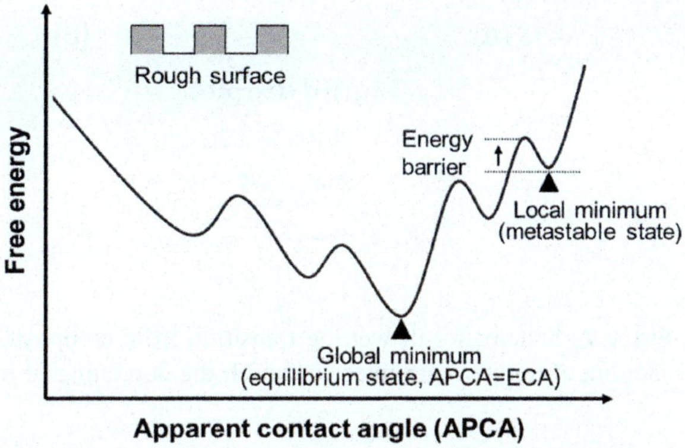

Fig. 1.6 Schematic of the change of free energy with APCA on rough surfaces

where r_W is the ratio of the actual to the projected solid-liquid interface area, which is $r_W = (l_1 + 2h)/l_1$ as shown in Fig. 1.5b given that the structure surface has negligible roughness. For an intrinsically hydrophobic or hydrophilic surface, the noncomposite state enhances its hydrophobicity or hydrophilicity since r_W is always greater than 1.

Figure 1.6 shows a schematic of the change of Gibbs free energy with APCA on a rough surface. The Cassie-Baxter equation and Wenzel equation give the ECA for composite and noncomposite state, respectively [27]. Different from Fig. 1.4, local minima of free energy exist in Fig. 1.6, so a droplet may be pinned at a metastable state and exhibit a metastable contact angle (MCA), because of free energy barrier [21, 28]. The largest and the smallest MCAs are advancing contact angle (ACA) and receding contact angle (RCA), respectively [2]. The difference between ACA and RCA is known as contact angle hysteresis (CAH).

1.2.3 Wetting Transition

Even if on the same rough surface, the wetting state of a liquid droplet may transit from the composite to noncomposite state upon receiving external perturbations, resulting in different ECAs and CAHs [9]. This transition process is termed wetting transition: liquid penetrates into the surface grooves, and finally the liquid-vapor interface under the droplet disappears. Wetting transition can be attributed to the lower system energy of the noncomposite state compared to the composite state. The two wetting states are separated by asymmetric free energy barriers, so wetting transition is generally considered to be irreversible [29]. When the composite state is energetically favorable, transition from noncomposite to composite state can be achieved [30, 31].

Figure 1.7 presents two possible mechanisms of wetting transition: (A) sagging and (B) depinning. For the sagging transition, the internal TPCLs remain pinned at the top edges of surface structures, while the liquid-vapor interface under droplet deforms and touches the groove bottom. The sagging transition is likely to occur on the surfaces with

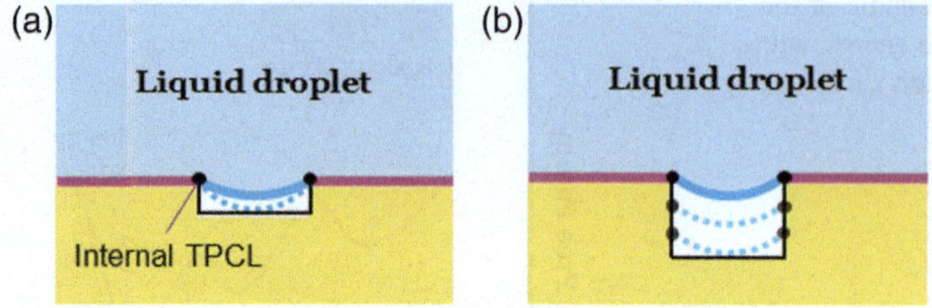

Fig. 1.7 Schematics of wetting transition from composite to noncomposite state because of **a** the sagging of liquid-vapor interface and **b** the depinning of internal TPCL [9]

low structure [32]. The depinning transition occurs when the droplet gains enough energy to overcome the free energy barrier between composite and noncomposite states. As a result, the internal TPCLs are de-pinned and proceed down the structure.

Composite state is usually featured with high contact angle and low CAH [9], which are needed for lotus superhydrophobic surfaces. Therefore, many efforts have been devoted to enhancing the stability of composite state by modifying surface structure and chemistry. For surface chemistry, post-treatment methods such as chemical or physical vapor deposition [33, 34] and plasma treatment [35] have been used to introduce chemicals or chemical groups to reduce surface energy. For surface structures, re-entrant structures such as micro hoodoo [36] and inverted trapezoid [37], and hierarchical structures [38] are developed to stabilize the metastable composite state. These surface structures can be fabricated by methods such as reactive ion etching [36], 3D diffuser lithography [37], or electrospinning [38].

1.3 Wetting Control by Electrospinning

There are a few methods, such as template synthesis, sol-gel method, laser/plasma/chemical etching, layer-by-layer deposition, and electrospinning [39, 40], to modify the wettability of a solid surface. Among them, electrospinning is a facile and versatile technique to produce continuous micro- and nano-fibers [41], providing adjustable surface roughness to control surface wettability [40]. Moreover, electrospinning is considered an additive manufacturing technology, allowing for surface modification on valuable substrate (e.g., for in-situ wound dressing [42]), and the introduced fibrous structure is featured with a high air permeability.

Figure 1.8 depicts a typical electrospinning setup that consists of three parts: a feeding device with a metallic needle, a high-voltage power supply, and a grounded collector [43]. The high-voltage power supply is connected to the needle, creating an electric field between the needle and grounded collector. As the polymer solution from the feeding device forms a droplet at the needle tip, surface charges are induced on the droplet, and

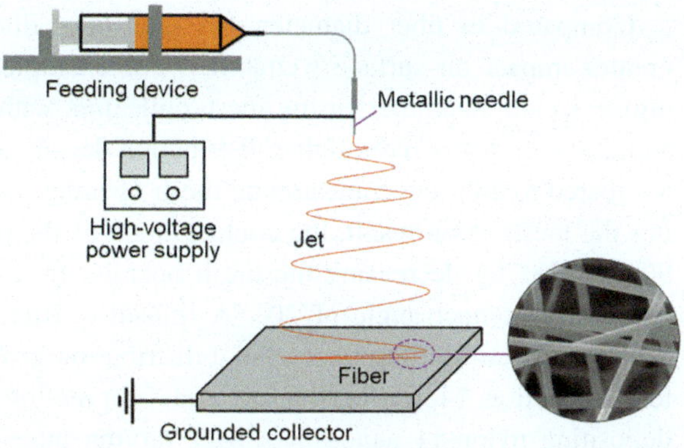

Fig. 1.8 Schematic of a typical electrospinning setup for fiber fabrication [43]

the electrostatic repulsion of surface charges deforms the droplet into a conical shape (also known as Taylor cone) [44]. Once the supply voltage exceeds a threshold value, the electrostatic repulsion overcomes the surface tension of droplet and a charged jet (fiber) is ejected from the droplet toward the collector. The charges on the fiber induce bending instability, causing stretching of the fiber and rapid evaporation of solvent [43]. At the end of the process, the fiber reaches the collector and non-woven mat is formed. It is worth noting that single-spinneret electrospinning may suffer from low throughput, resulting in a low production rate. This issue can be addressed by multiple-needle electrospinning [45].

Surface wetting is controlled by both surface structure and surface energy. Based on this principle, modifying surface wettability using electrospun fibers is realized by controlling fiber morphology and/or fiber surface chemistry, which are discussed in detail as follows.

1.3.1 Modifying Surface Wettability by Controlling Fiber Morphology

Based on surface morphology, the electrospun fibers can be categorized into (1) fibers with smooth surface and (2) fibers with secondary morphology including beads, pores, grooves, and roughness [46]. Droplets on these two types of fibers have different wetting behaviors, so they are discussed separately.

For ordinary fibers with relatively smooth surface, the diameter, inter-fiber distance, and spatial distribution of fibers can influence surface wettability. For example, Ma et al. [47] fabricated randomly distributed poly(caprolactone) (PCL) fibers with various diameters by adjusting the PCL solution concentration for electrospinning; water contact angle increases from 119° to 129° as fiber diameter decreases from 2.2 to 0.58 μm. The increase of contact angle was attributed to the higher surface roughness by thinner fibers.

Compared to fiber diameter, the inter-fiber distance and spatial distribution show greater impact on surface wettability. For example, Kakunuri et al. [48] fabricate cellulose acetate nanofibers using mesh collectors with various opening sizes, i.e., 50, 100, and 200 μm; fibers have higher distribution density around mesh lines, i.e., 77% coverage compared to 28% coverage around mesh openings, so the fibrous mat is patterned following the mesh. As a result, the contact angle of the patterned fibrous mats increases from 90° to 138° by decreasing the mesh opening from 200 to 50 μm, while non-patterned mat has a contact angle of 30°. A transition from hydrophilicity to hydrophobicity is achieved by controlling fiber spatial distribution given that the fibers have similar diameter distribution. Moreover, Bagrov et al. [49] modifies the wettability of aluminum foil by depositing nylon-11 nano-fibers with varying inter-fiber distance. Surface coverage rate of fibers increases from 0.7 to 20%, where the inter-fiber pore size is around 2.5 μm, by increasing electrospinning time from 2 to 5 min. As a result, contact angle increases from ~70° to 124°.

The large variation of contact angle in Refs. [48, 49] is explained by the change of wetting state. Specifically, droplets on fibers-dressed surfaces tend to change from Wenzel state to Cassie-Baxter state with decreasing inter-fiber distance, leading to larger equilibrium contact angles. However, the influence of fiber diameter on wetting state is relatively small due to the limited adjustable range of fiber diameter by electrospinning.

Based on electrospinning, fibers with various secondary morphologies can be generated. Figure 1.9 shows the scanning electron microscope (SEM) images of (A) beaded [50], (B) porous, (C) grooved [51], and (D) wrinkled [52] fibers. These secondary morphologies are proven to have significant influences on the wetting properties of fibers.

Beaded fiber, or bead-on-string structure, which can be produced by reducing the applied voltage or the concentration of the precursor solution for electrospinning [53], can enhance the surface hydrophobicity. For example, Hong et al. [54] prepared beaded poly(vinylidene fluoride) (PVDF) fibers by reducing the concentration of PVDF solution from 25 wt%, which is for beadless fibers, to 20 wt%. The beaded fibers are featured with 5–9 μm beads and 100–150 nm interconnected fibers, while the diameter of beadless fibers is around 960 nm. As a result, the contact angle increases from 128° to 150° [54]. The reason lies in that the large beads, as protrusions of fibrous mat, hold droplet with air pockets underneath. However, the beaded fibers also show high CAH, indicating that there is contact between droplet and fibers along TPCL.

As another example, Wong et al. [38] used polystyrene (PS) solutions with concentration of 8 wt%, 10 wt%, and 20 wt% for electrospinning, resulting in heavily beaded fibers (~650 beads/mm^2), partially beaded fibers (~160 beads/mm^2), and beadless fibers, respectively. The average bead diameter is 5.0 μm and 13.5 μm for heavily and partially beaded fibers, respectively. It turns out that contact angle increases from ~140° for beadless fibers to ~150° and ~160° for partially and heavily beaded fibers, respectively. In addition, droplets with a volume of 5 μL can roll off from the surfaces covered by beaded fibers with roll-off angles of 28°–46°.

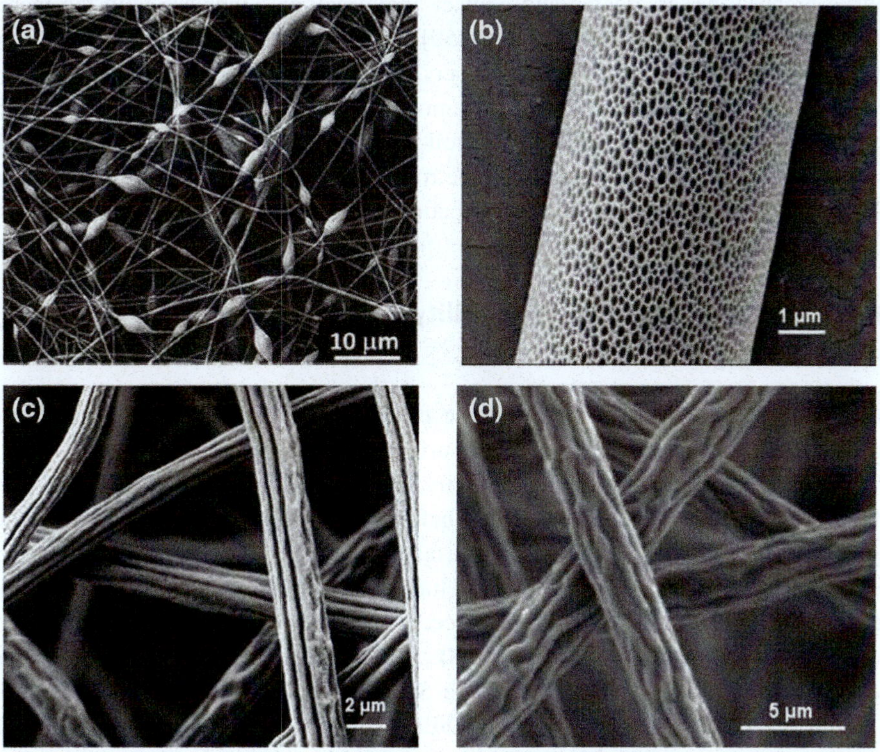

Fig. 1.9 Scanning electron microscope image of **a** beaded [50], **b** porous, **c** grooved [51], and **d** wrinkled [52] electrospun fibers

Porous fibers can be fabricated by (1) phase separation of solution jets in electrospinning, (2) condensation of water vapor onto the jets in electrospinning, and (3) selective removal of sacrificial component after electrospinning [55]. For porous fibers, a composite of solid-liquid and liquid-air interfaces may be formed at fiber surface, resulting in an increment of hydrophobicity [34]. For example, Miyauchi et al. [56] fabricated porous PS fibers based on the phase separation of solution jet in electrospinning; the solvent to prepare PS solution is a mixture of N,N-dimethylformamide (DMF) and tetrahydrofuran (THF). As a result, the pores on the fiber surface increase contact angle from 144°, for smooth fibers, to 158°, and decrease roll-off angle for 12 μL droplet by 25°. Moreover, McCann et al. [57] and Ma et al. [34] have proven that porous PVDF and poly(methyl methacrylate) PMMA, respectively, fibers can also increase contact angle compared to nonporous fibers.

Different from porous fibers, a noncomposite solid-liquid contact interface may form at fiber surface for grooved or wrinkled fibers because the grooves or wrinkles may be wetted by liquid for their low structure height and large opening area [58, 59]. As a result, a higher CAH is created and contact angle may also increase with the increased fiber roughness. For instance, Liang et al. [59] compared the wetting behaviors of grooved and smooth poly(L-lactic acid) (PLLA) fibers. The grooved fibers have a contact angle of 145°, compared with that of 135° for the smooth fibers. CAHs are compared by aligning fibers and measuring the difference of the roll-off angles along two directions: parallel and normal to the alignment. Consequently, the roll-off angle difference for grooved fibers is 4° higher than that for smooth fibers, indicating a higher CAH.

1.3.2 Modifying Surface Wettability by Controlling Fiber Surface Energy

Fiber surface energy is determined by the chemical compositions at the fiber surface. There are mainly two ways to control the surface chemistry: (1) adjust precursor solution before electrospinning, and (2) apply post treatment after electrospinning.

Polymer blending in precursor solution before electrospinning is a simple and efficient way to modify the wettability of fibrous surfaces, especially in creating superhydrophilicity. Usually, two polymers with opposite wettability, i.e., hydrophobic or weak hydrophilic and hydrophilic, are blended. For instance, Kurusu et al. [60] blended poly(styrene-*b*-(ethylene-*co*-butylene)-*b*-styrene) (SEBS) with Pluronic F127 to turn hydrophobic electrospun SEBS fibers hydrophilic. Pure SEBS fibers have a contact angle of 139°. By adding 15–20 wt% Pluronic F127 into SEBS solution, the obtained SEBS/F127 fibers show a contact angle of 0°. Moreover, Li et al. [61] blended hydrophobic poly(ethylene terephthalate) (PET) and hydrophilic poly(vinyl alcohol) (PVA) in hexafluoro isopropanol (HFIP) for electrospinning. Compared with the contact angle of 131° for pure PET fibers, the obtained PET/PVA fibers show 0° contact angle even when a small amount of PVA is used, i.e., 1/20 weight of PET. However, for the polymer blending method, the change of contact angle with blending ratio is not continuous because the hydrophilic polymer tends to aggregate at fiber surface [60, 61].

In addition to the polymer blending, the precursor solution can also be adjusted by adding inorganics. In contrast with that by polymer blending, the change of contact angle by adding inorganics can be continuous. For example, Abdal-hay et al. [62] added hydroxyapatite (Hap) into nylon-6 solution for electrospinning. The nylon-6/Hap composite fibers show contact angles continuously varying from 130° to 0° with increasing Hap from 0 to 10 wt%. This continuous change of contact angle can also be achieved by adding Triton X-100 non-ionic surfactant into polyamide 6 (PA6) solution [63].

Compared to adjusting precursor solution, applying post treatments after electrospinning to modify fiber surface chemistry has less restriction on applicable polymers, electrospinning parameters, etc. A few post-treatment methods have been used, including initiated chemical vapor deposition (iCVD) [34, 47], plasma etching [54], dip coating [64, 65]. For example, Ma et al. [47] used iCVD to deposit poly(perfluoroalkyl ethyl methacrylate) (PPFEMA), a low surface energy (0.0093 J/m^2) polymer, onto electrospun PCL fibers. After iCVD, the contact angle of PCL fibers increases from 120°–130° to 150°–160°, and 20 μL droplets can roll off the surface at threshold tilting angles of 6°–12°. As another example, Schoolaert et al. [64] modified poly(2-n-propyl-2-oxazoline) (PnPrOx) nanofibers with diameter of 200–400 nm by dip-coating into tannic acid. By increasing tannic acid concentration from 0 to 10 wt%, contact angle decreases from 120° to 40°. It should be noted that many relevant works also combine the modification of fiber morphology as discussed later.

1.3.3 Modifying Surface Wettability by Combining Fiber Morphology and Surface Energy

Controlling both fiber morphology and surface energy is usually aimed at creating extreme wettability, i.e., superhydrophilicity or superhydrophobicity. To develop superhydrophilic fibrous surfaces, the basic principle is to increase fiber surface energy on a rough surface. As explained by Wenzel's equation, Eq. (1.3), roughness intensifies the hydrophilicity for intrinsic hydrophilic materials.

To increase fiber surface energy, many scholars add inorganics such as metal oxides [66, 67] and silicate minerals [68] into precursor solution, followed by thermal treatment after electrospinning to expose inorganics at fiber surfaces. For example, Wang et al. [66] added $Ti(Obu)_4$ into poly(vinyl pyrrolidone) (PVP) solution for electrospinning. After calcination of as-spun fibers under 500 °C, TiO_2 nanofibers with nanochannels on surfaces are obtained. The contact angle of 2 μL droplet reaches 0° within 30 ms after droplet deposition. Similarly, Zhu et al. [67] mixed $FeAc_2$ and PVA for precursor solution, and α-Fe_2O_3 nanofibers with rough surfaces were obtained after calcination. As a result, droplets on the α-Fe_2O_3 fibrous surfaces spread to 0° contact angle within hundreds of milliseconds. Moreover, Tijing et al. [68] decorated hydrophobic polyurethane (PU) fibers with hydrophilic tourmaline I nano-particles by adding TM nanoparticles into PU precursor solution. Without the need of thermal treatment, the as-spun beaded fibers mats have contact angles varying from 125° to 13° by increasing TM amount from 0 to 5 wt%.

1.4 Opportunities and Challenges

As a result, using electrospinning to develop lotus or rose-petal superhydrophobicity gains growing attention in recent years, which can also be evidenced by the rising numbers of the annual publications, as shown in Fig. 1.10.

The electrospinning-based methods for developing superhydrophobicity, which will be discussed in detail in Chap. 2, can be classified into one-step electrospinning and

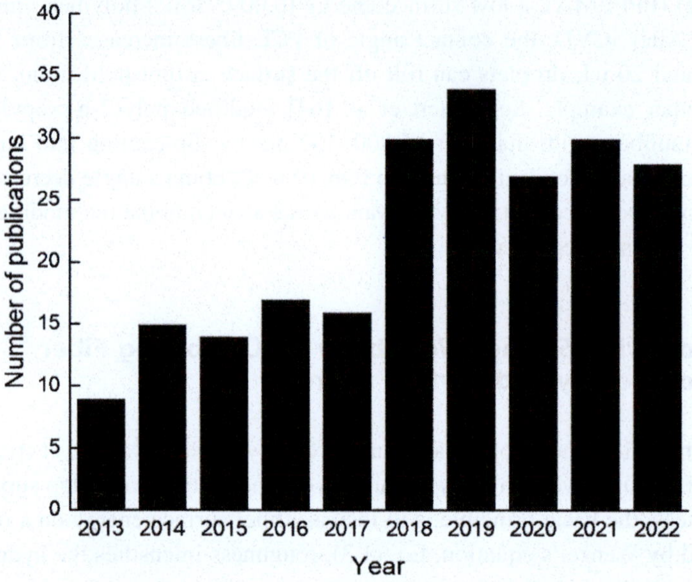

Fig. 1.10 Number of publications in creating superhydrophobicity by electrospinning (data obtained by searching keywords "electrospinning" and "superhydrophobicity" from Web of Science database on April 28, 2023)

electrospinning followed by post treatments. For the one-step electrospinning methods, superhydrophobicity is directly realized on the as-spun fibrous structures. For methods with post treatments, superhydrophobicity is achieved by modifications of the fibrous surfaces after electrospinning.

At present, most electrospinning methods realize superhydrophobicity by increasing fiber surface roughness and/or reducing fiber surface energy. To increase fiber surface roughness, a secondary fiber morphology including beads [69] and nanoparticles [70] is often created on the fiber surfaces in electrospinning [71] or after electrospinning by post treatments [72]. Moreover, methods such as copolymer [73] and coaxial [74] electrospinning of a low surface energy polymer with an electrospinnable polymer, and coating the as-spun fibers with a thin layer of a low surface energy material such as poly(perfluoroalkyl ethyl methacrylate) [34] after electrospinning have been reported to reduce the fiber surface energy.

However, a strategic approach to superhydrophobicity in electrospun fibers is still lacking. Current theoretical works have limited abilities in guiding the development of fiber-based superhydrophobicity because either the works are limited to only investigating the stability of wetting state or the theoretical structures are difficult to replicate. The control of fiber morphology parameters, including fiber diameter, inter-fiber distance, and fiber orientation, in previous works has often relied on trial-and-error approaches.

As a result, thick fibrous mats are usually made to reduce the heterogeneity of fiber morphology, reducing the air permeability and transparency. In addition, the fibers may show different wettability despite similar intrinsic contact angles and secondary fiber morphologies [54, 69, 75].

Moreover, existing electrospinning-based methods usually involve complex fabrication processes for achieving superhydrophobicity. To induce secondary fiber morphologies, for example, nanoparticles are blended with polymer solutions for electrospinning or coated on the fibers after electrospinning. Homogeneous mixing of nanoparticles with a polymer solution takes extra efforts to disperse the nanoparticles [76], and post treatments are needed to coat fibers with nanoparticles. Alternatively, copolymer and coaxial electrospinning, and coating fiber surface after electrospinning are often adopted to reduce fiber surface energy. All these can complicate the fabrication and limit their applications. Therefore, a simple one-step electrospinning method without additive is needed to develop superhydrophobicity.

This book aims to develop simple electrospinning-based methods for the fabrication of petal and lotus superhydrophobic surfaces with rational approach. The objective is achieved by the following steps. First, understanding droplet wetting behavior on parallel-structured surfaces helps to develop superhydrophobicity. The influence of fiber morphology on the wettability is modeled and experimentally validated. Then, superhydrophobic surfaces with petal effect are developed using aligned fibers fabricated by one-step electrospinning. After that, the wetting behavior of droplets on uniaxially (mono layer) and biaxially (dual layers) aligned porous fibers is studied. Later, lotus surfaces are fabricated using biaxially aligned porous fibers made by one-step electrospinning. Finally, the developed lotus surface is used to develop a waterproof, breathable, and superhydrophobic membrane, and a surface layer for wearable electronics.

Accordingly, the book is organized as follows. Chapter 2 provides a detailed literature review about developing superhydrophobicity based on electrospinning is provided. For both lotus and rose-petal superhydrophobicity, the electrospinning-based methods are classified into one-step electrospinning and electrospinning followed by post treatments based on the fabrication steps.

Chapter 3 details an integrated thermodynamic model for droplets wetting on parallel-structured surfaces with various cross-sections, including circle, elongated circle, trapezoid, rectangle, and inverted trapezoid. The model relates free energy to the continuous movement of three-phase contact lines in different wetting states, and the forming conditions for free energy barrier are specified. Then, the model and the forming conditions of energy barrier are validated experimentally by electrospun fibers and existing experimental and theoretical works. Based on the validated model, the wetting behaviors on various parallel-structured surfaces are studied and compared.

Chapter 4 introduces the development of petal-effect surfaces based on uniaxially aligned electrospun fibers. First, aligned polystyrene fibers with various fiber diameters and inter-fiber distances are produced onto substrates by simple one-step electrospinning, followed by characterization of fiber morphology and wettability. The influence of fiber morphology parameters on the wettability of surfaces dressed by aligned fibers is experimentally studied with explanation of mechanism based on a wetting model for aligned fibers. The model is specified from the integrated model introduced in Chap. 3. Then petal-effect surfaces are constructed using aligned fibers based on the influence of fiber parameters.

Chapter 5 adopts aligned porous fibers to develop lotus effect because the porous fiber surface reduces the contact area between fibers and droplets and the alignment serves as a track to facilitate droplets rolling off. First, uniaxially and biaxially aligned porous fibers are deposited onto substrate surfaces by one-step electrospinning with controlled relative humidity. Then the measured hydrophobicity of surfaces dressed by uniaxially and biaxially aligned porous fibers is compared and explained by numerical modeling. A numerical model is developed in Surface Evolver to determine the wetting state of droplets on fiber-dressed surfaces. Finally, based on the biaxially aligned porous fibers, lotus surfaces are constructed following a strategic approach considering the influence of porous fiber parameters.

Chapter 6 demonstrates the potential engineering applications of the biaxial structure in Chap. 5 based on the lotus effect and the expected high breathability, only consisting of two layers of aligned fibers. Two engineering applications are demonstrated, including (1) development of waterproof, breathable, and superhydrophobic membrane, and (2) a surface layer for protecting wearable electronics.

Chapter 7 summarizes the conclusions of this book and then recommends the research directions for the future.

References

1. A.R. Parker, C.R. Lawrence, Water capture by a desert beetle. Nature **414**(6859), 33–34 (2001)
2. J. Long, M. Hyder, R. Huang, P. Chen, Thermodynamic modeling of contact angles on rough, heterogeneous surfaces. Adv. Coll. Interface. Sci. **118**(1–3), 173–190 (2005)
3. C. Jothi Prakash, R. Prasanth,Approaches to design a surface with tunable wettability: a review on surface properties. J. Mater. Sci. **56**, 108–135 (2021)
4. Z. Wang, M. Elimelech, S. Lin, Environmental applications of interfacial materials with special wettability. Environ. Sci. Technol. **50**(5), 2132–2150 (2016)
5. L. Sun, J. Guo, H. Chen, D. Zhang, L. Shang, B. Zhang et al., Tailoring materials with specific wettability in biomedical engineering. Adv. Sci. **8**(19), 2100126 (2021)
6. J. Drelich, E. Chibowski, Superhydrophilic and superwetting surfaces: definition and mechanisms of control. Langmuir **26**(24), 18621–18623 (2010)
7. X. Deng, L. Mammen, H.-J. Butt, D. Vollmer, Candle soot as a template for a transparent robust superamphiphobic coating. Science **335**(6064), 67–70 (2012)
8. D. Ebert, B. Bhushan, Wear-resistant rose petal-effect surfaces with superhydrophobicity and high droplet adhesion using hydrophobic and hydrophilic nanoparticles. J. Colloid Interface Sci. **384**(1), 182–188 (2012)

9. E. Bormashenko, Progress in understanding wetting transitions on rough surfaces. Adv. Coll. Interface. Sci. **222**, 92–103 (2015)
10. W. Barthlott, C. Neinhuis, Purity of the sacred lotus, or escape from contamination in biological surfaces. Planta **202**, 1–8 (1997)
11. L. Feng, Y. Zhang, J. Xi, Y. Zhu, N. Wang, F. Xia et al., Petal effect: a superhydrophobic state with high adhesive force. Langmuir **24**(8), 4114–4119 (2008)
12. https://commons.wikimedia.org/w/index.php?curid=364500
13. Y. Zheng, C. Zhang, J. Wang, Y. Liu, C. Shen, J. Yang, Robust adhesion of droplets via heterogeneous dynamic petal effects. J. Colloid Interface Sci. **557**, 737–745 (2019)
14. K. Liu, L. Jiang, Bio-inspired self-cleaning surfaces. Annu. Rev. Mater. Res. **42**, 231–263 (2012)
15. B. Zhang, X. Zhao, Y. Li, B. Hou, Fabrication of durable anticorrosion superhydrophobic surfaces on aluminum substrates via a facile one-step electrodeposition approach. RSC Adv. **6**(42), 35455–35465 (2016)
16. T.V. Charpentier, A. Neville, P. Millner, R.W. Hewson, A. Morina, Development of anti-icing materials by chemical tailoring of hydrophobic textured metallic surfaces. J. Colloid Interface Sci. **394**, 539–544 (2013)
17. Y. Chen, Y. Zhang, L. Shi, J. Li, Y. Xin, T. Yang et al., Transparent superhydrophobic/superhydrophilic coatings for self-cleaning and anti-fogging. Appl. Phys. Lett. **101**(3), 033701 (2012)
18. X. Hong, X. Gao, L. Jiang, Application of superhydrophobic surface with high adhesive force in no lost transport of superparamagnetic microdroplet. J. Am. Chem. Soc. **129**(6), 1478–1479 (2007)
19. A.I. Neto, C.R. Correia, C.A. Custódio, J.F. Mano, Biomimetic miniaturized platform able to sustain arrays of liquid droplets for high-throughput combinatorial tests. Adv. Func. Mater. **24**(32), 5096–5103 (2014)
20. A. Winkleman, G. Gotesman, A. Yoffe, R. Naaman, Immobilizing a drop of water: fabricating highly hydrophobic surfaces that pin water droplets. Nano Lett. **8**(4), 1241–1245 (2008)
21. A. Marmur, Soft contact: measurement and interpretation of contact angles. Soft Matter **2**(1), 12–17 (2006)
22. T. Young, III. An essay on the cohesion of fluids. Philos. Trans. R. Soc. Lond. **95**, 65–87 (1805)
23. P.-G. Gennes, F. Brochard-Wyart, D. Quéré, *Capillarity and Wetting Phenomena: drops, Bubbles, Pearls, Waves* (Springer, 2004)
24. A. Marmur, Thermodynamic aspects of contact angle hysteresis. Adv. Coll. Interface. Sci. **50**, 121–141 (1994)
25. A. Cassie, S. Baxter, Wettability of porous surfaces. Trans. Faraday Soc. **40**, 546–551 (1944)
26. R.N. Wenzel, Resistance of solid surfaces to wetting by water. Ind. Eng. Chem. **28**(8), 988–994 (1936)
27. A. Marmur, Wetting on hydrophobic rough surfaces: to be heterogeneous or not to be? Langmuir **19**(20), 8343–8348 (2003)
28. C.H. Kung, P.K. Sow, B. Zahiri, W. Mérida, Assessment and interpretation of surface wettability based on sessile droplet contact angle measurement: challenges and opportunities. Adv. Mater. Interfaces **6**(18), 1900839 (2019)
29. G. Whyman, E. Bormashenko, Wetting transitions on rough substrates: general considerations. J. Adhes. Sci. Technol. **26**(1–3), 207–220 (2012)
30. J.B. Boreyko, C.-H. Chen, Restoring superhydrophobicity of lotus leaves with vibration-induced dewetting. Phys. Rev. Lett. **103**(17), 174502 (2009)
31. R.J. Vrancken, H. Kusumaatmaja, K. Hermans, A.M. Prenen, O. Pierre-Louis, C.W. Bastiaansen et al., Fully reversible transition from Wenzel to Cassie-Baxter states on corrugated superhydrophobic surfaces. Langmuir **26**(5), 3335–3341 (2010)

32. J.R. Panter, H. Kusumaatmaja, The impact of surface geometry, cavitation, and condensation on wetting transitions: posts and reentrant structures. J. Phys.: Condens. Matter **29**(8), 084001 (2017)
33. L. Huang, S.P. Lau, H. Yang, E. Leong, S.F. Yu, S. Prawer, Stable superhydrophobic surface via carbon nanotubes coated with a ZnO thin film. J. Phys. Chem. B **109**(16), 7746–7748 (2005)
34. M. Ma, M. Gupta, Z. Li, L. Zhai, K.K. Gleason, R.E. Cohen et al., Decorated electrospun fibers exhibiting superhydrophobicity. Adv. Mater. **19**(2), 255–259 (2007)
35. Y.I. Yoon, H.S. Moon, W.S. Lyoo, T.S. Lee, W.H. Park, Superhydrophobicity of PHBV fibrous surface with bead-on-string structure. J. Colloid Interface Sci. **320**(1), 91–95 (2008)
36. A. Tuteja, W. Choi, J.M. Mabry, G.H. McKinley, R.E. Cohen, Robust omniphobic surfaces. Proc. Natl. Acad. Sci. **105**(47), 18200–18205 (2008)
37. M. Im, H. Im, J.-H. Lee, J.-B. Yoon, Y.-K. Choi, A robust superhydrophobic and superoleophobic surface with inverse-trapezoidal microstructures on a large transparent flexible substrate. Soft Matter **6**(7), 1401–1404 (2010)
38. W.S. Wong, N. Nasiri, G. Liu, N. Rumsey-Hill, V.S. Craig, D.R. Nisbet et al., Flexible transparent hierarchical nanomesh for rose petal-like droplet manipulation and lossless transfer. Adv. Mater. Interfaces **2**(9), 1500071 (2015)
39. N. Nuraje, W.S. Khan, Y. Lei, M. Ceylan, R. Asmatulu, Superhydrophobic electrospun nanofibers. J. Mater. Chem. A **1**(6), 1929–1946 (2013)
40. M. Ma, R.M. Hill, Superhydrophobic surfaces. Curr. Opin. Colloid Interface Sci. **11**(4), 193–202 (2006)
41. T. Subbiah, G.S. Bhat, R.W. Tock, S. Parameswaran, S.S. Ramkumar, Electrospinning of nanofibers. J. Appl. Polym. Sci. **96**(2), 557–569 (2005)
42. Y. Yue, X. Gong, W. Jiao, Y. Li, X. Yin, Y. Si et al., In-situ electrospinning of thymol-loaded polyurethane fibrous membranes for waterproof, breathable, and antibacterial wound dressing application. J. Colloid Interface Sci. **592**, 310–318 (2021)
43. F.K. Ko, Y. Wan, Nanofiber technology, in *Introduction to Nanofiber Materials* (Cambridge University Press, 2014), pp. 44–63
44. G.I. Taylor. Electrically driven jets. Proc. R. Soc. London. A. Math. Phys. Sci. **313**(1515), 453–475 (1969).
45. J. Xue, T. Wu, Y. Dai, Y. Xia, Electrospinning and electrospun nanofibers: methods, materials, and applications. Chem. Rev. **119**(8), 5298–5415 (2019)
46. B. Zaarour, L. Zhu, X. Jin, A review on the secondary surface morphology of electrospun nanofibers: formation mechanisms, characterizations, and applications. ChemistrySelect **5**(4), 1335–1348 (2020)
47. M. Ma, Y. Mao, M. Gupta, K.K. Gleason, G.C. Rutledge, Superhydrophobic fabrics produced by electrospinning and chemical vapor deposition. Macromolecules **38**(23), 9742–9748 (2005)
48. M. Kakunuri, N.D. Wanasekara, C.S. Sharma, M. Khandelwal, S.J. Eichhorn,Three-dimensional electrospun micropatterned cellulose acetate nanofiber surfaces with tunable wettability. J. Appl. Polym. Sci. **134**(15) (2017)
49. D. Bagrov, S. Perunova, E. Pavlova, D. Klinov, Wetting of electrospun nylon-11 fibers and mats. RSC Adv. **11**(19), 11373–11379 (2021)
50. N. Thakur, A.S. Ranganath, K. Agarwal, A. Baji, Electrospun bead-on-string hierarchical fibers for fog harvesting application. Macromol. Mater. Eng. **302**(7), 1700124 (2017)
51. C. Huang, H. Niu, J. Wu, Q. Ke, X. Mo, T. Lin, Needleless electrospinning of polystyrene fibers with an oriented surface line texture. J. Nanomater. **2012**, 1–7 (2012)
52. C.-L. Pai, M.C. Boyce, G.C. Rutledge, Morphology of porous and wrinkled fibers of polystyrene electrospun from dimethylformamide. Macromolecules **42**(6), 2102–2114 (2009)

53. Z. Asvar, E. Mirzaei, N. Azarpira, B. Geramizadeh, M. Fadaie, Evaluation of electrospinning parameters on the tensile strength and suture retention strength of polycaprolactone nanofibrous scaffolds through surface response methodology. J. Mech. Behav. Biomed. Mater. **75**, 369–378 (2017)
54. H.R. Hong, C.H. Park, The influence of nanostructure on the wetting transition of polyvinylidene fluoride nanoweb: from the petal effect to the lotus effect. Text. Res. J. **91**(7–8), 752–765 (2021)
55. C. Huang, N.L. Thomas, Fabrication of porous fibers via electrospinning: strategies and applications. Polym. Rev. **60**(4), 595–647 (2020)
56. Y. Miyauchi, B. Ding, S. Shiratori, Fabrication of a silver-ragwort-leaf-like super-hydrophobic micro/nanoporous fibrous mat surface by electrospinning. Nanotechnology **17**(20), 5151 (2006)
57. J.T. McCann, M. Marquez, Y. Xia, Highly porous fibers by electrospinning into a cryogenic liquid. J. Am. Chem. Soc. **128**(5), 1436–1437 (2006)
58. A. Rawal, S. Shukla, S. Sharma, D. Singh, Y.-M. Lin, J. Hao et al., Metastable wetting model of electrospun mats with wrinkled fibers. Appl. Surf. Sci. **551**, 149147 (2021)
59. M. Liang, X. Chen, Y. Xu, L. Zhu, X. Jin, C. Huang, Double-grooved nanofibre surfaces with enhanced anisotropic hydrophobicity. Nanoscale **9**(42), 16214–16222 (2017)
60. R.S. Kurusu, N.R. Demarquette, Blending and morphology control to turn hydrophobic SEBS electrospun mats superhydrophilic. Langmuir **31**(19), 5495–5503 (2015)
61. G. Li, Y. Zhao, M. Lv, Y. Shi, D. Cao, Super hydrophilic poly (ethylene terephthalate)(PET)/poly (vinyl alcohol)(PVA) composite fibrous mats with improved mechanical properties prepared via electrospinning process. Colloids Surf. A **436**, 417–424 (2013)
62. A. Abdal-Hay, H.R. Pant, J.K. Lim, Super-hydrophilic electrospun nylon-6/hydroxyapatite membrane for bone tissue engineering. Eur. Polym. J. **49**(6), 1314–1321 (2013)
63. R. Beigmoradi, A. Samimi, D. Mohebbi-Kalhori, Fabrication of polymeric nanofibrous mats with controllable structure and enhanced wetting behavior using one-step electrospinning. Polymer **143**, 271–280 (2018)
64. E. Schoolaert, L. Cossu, J. Becelaere, J.F. Van Guyse, A. Tigrine, M. Vergaelen et al., Nanofibers with a tunable wettability by electrospinning and physical crosslinking of poly (2-n-propyl-2-oxazoline). Mater. Des. **192**, 108747 (2020)
65. T. Ogawa, B. Ding, Y. Sone, S. Shiratori, Super-hydrophobic surfaces of layer-by-layer structured film-coated electrospun nanofibrous membranes. Nanotechnology **18**(16), 165607 (2007)
66. L. Wang, Y. Zhao, J. Wang, X. Hong, J. Zhai, L. Jiang et al., Ultra-fast spreading on superhydrophilic fibrous mesh with nanochannels. Appl. Surf. Sci. **255**(9), 4944–4949 (2009)
67. Y. Zhu, J.C. Zhang, J. Zhai, L. Jiang, Preparation of superhydrophilic α-Fe2O3 nanofibers with tunable magnetic properties. Thin Solid Films **510**(1–2), 271–274 (2006)
68. L.D. Tijing, M.T.G. Ruelo, A. Amarjargal, H.R. Pant, C.-H. Park, D.W. Kim et al., Antibacterial and superhydrophilic electrospun polyurethane nanocomposite fibers containing tourmaline nanoparticles. Chem. Eng. J. **197**, 41–48 (2012)
69. L. Jiang, Y. Zhao, J. Zhai, A lotus-leaf-like superhydrophobic surface: a porous microsphere/nanofiber composite film prepared by electrohydrodynamics. Angew. Chem.-Int. Ed. **43**(33), 4338–4341 (2004)
70. J. Lin, Y. Cai, X. Wang, B. Ding, J. Yu, M. Wang, Fabrication of biomimetic superhydrophobic surfaces inspired by lotus leaf and silver ragwort leaf. Nanoscale **3**(3), 1258–1262 (2011)
71. S. Jin, Y. Park, C.H. Park, Preparation of breathable and superhydrophobic polyurethane electrospun webs with silica nanoparticles. Text. Res. J. **86**(17), 1816–1827 (2016)
72. H. Gu, G. Li, P. Li, H. Liu, T.T. Chadyagondo, N. Li et al., Superhydrophobic and breathable SiO2/polyurethane porous membrane for durable water repellent application and oil-water separation. Appl. Surf. Sci. **512**, 144837 (2020)

73. M. Ma, R.M. Hill, J.L. Lowery, S.V. Fridrikh, G.C. Rutledge, Electrospun poly (styrene-block-dimethylsiloxane) block copolymer fibers exhibiting superhydrophobicity. Langmuir **21**(12), 5549–5554 (2005)
74. D. Han, A.J. Steckl, Superhydrophobic and oleophobic fibers by coaxial electrospinning. Langmuir **25**(16), 9454–9462 (2009)
75. J. Serra, R. Pinto, J. Barbosa, D. Correia, R. Gonçalves, M. Silva et al., Ionic liquid based Fluoropolymer solid electrolytes for Lithium-ion batteries. Sustain. Mater. Technol. **25**, e00176 (2020)
76. X. Li, X. Yu, C. Cheng, L. Deng, M. Wang, X. Wang, Electrospun superhydrophobic organic/inorganic composite nanofibrous membranes for membrane distillation. ACS Appl. Mater. Interfaces **7**(39), 21919–21930 (2015)

2 Methods to Develop Superhydrophobicity by Electrospinning

Electrospinning is proven an efficient method for fabricating superhydrophobic surfaces. In this section, the electrospinning-based methods for fabricating superhydrophobic surfaces are introduced, i.e., Sects. 2.1 and 2.2 for lotus superhydrophobicity and Sect. 2.3 for rose-petal superhydrophobicity. Based on the fabrication steps, the methods are classified into (1) one-step electrospinning and (2) electrospinning followed by post treatment.

2.1 One-Step Electrospinning

One-step electrospinning refers to that superhydrophobic surfaces can be directly obtained after electrospinning without any post treatment. Table 2.1 summarizes the one-step methods in literature. The first two columns explain the mechanism and materials, respectively, to achieve superhydrophobicity, and the third and fourth columns show the measured surface wettability. The mechanism is essentially controlling fiber morphology or/and fiber surface energy, as discussed in Sect. 1.3.

As shown in Table 2.1, most one-step methods rely on fabricating fibers with secondary morphology, especially beaded fibers, to realize lotus superhydrophobicity. A hydrophobic or weak hydrophilic polymer such as PS [1–3] and PCL [5] is often used to prepare a precursor solution that has a lower concentration than that for beadless fibers. As a result, beaded fibers with microbeads and interconnected nanofibers are obtained. When droplets are deposited on the surface, the intrinsically hydrophobic or weak hydrophilic microbeads with higher height than nanofibers can support the droplets in Cassie-Baxter state, resulting in lotus superhydrophobicity. For example, Jiang et al. [1] fabricated beaded PS fibers using 7 wt% PS/DMF solution. The diameter of beads and fibers is 3–7

Y. Zhang, *Superhydrophobicity of Surfaces Dressed by Electrospun Fibers*, Synthesis Lectures on Green Energy and Technology, https://doi.org/10.1007/978-3-031-55552-7_2

Table 2.1 One-step electrospinning to develop lotus superhydrophobicity

Mechanism	Materials	Contact angle	Roll-off angle (droplet volume)	Refs.
Beaded fibers	PS	160°	–	[1]
Beaded fibers	PANI/PS	162°	<5°	[2]
Beaded fibers	PS	155°	5°	[3]
Beaded fibers	PVDF-HFP	157°	6° (10 μL)	[4]
Beaded fibers doped with TiO_2 NPs	PCL/TiO_2	156°	–	[5]
Fibers with nano-protrusions	PS	154°	–	[6]
Fibers doped with TiO_2 NPs or graphene NFs	PVC/TiO_2 PS/TiO_2 PVC/graphene	168° 178° 166°	–	[7]
Fibers doped with SiO_2 NPs	PS/SiO_2	157°	2.2°[a]	[8]
Patterned fibers	PS	167°	5° (5 μL)	[9]
Copolymer fibers	PS-PDMS/PS	163°	15°[a]	[10]
Grooved fibers with Teflon outer layer	PCL/Teflon	158°	7° (10 μL)	[11]

PANI: polyaniline; PVDF-HFP: poly(vinylidene fluoride-co-hexafluoropropylene); PVC: polyvinyl chloride; PDMS: poly(methylhydrosiloxane); NPs: nanoparticles; NFs: nanoflakes
[a] Contact angle hysteresis

μm and 60–140 nm, respectively. The bead-on-string structure increases contact angle to 160°, compared to 139° for beadless fibers of 420 nm by 25 wt% PS/DMF solution. Based on Cassie-Baxter model, Eq. (1.2), the fraction of liquid-air interface is calculated to be 0.934 [1]. However, beaded fibers have high structural heterogeneity, resulting in a low mechanical strength [12] or even being hard to be free-standing [1].

Another common secondary fiber morphology in developing superhydrophobicity is the nano-protrusions on fiber surface, which are created by adding nano-particles [7, 8] or flakes [7] into the polymer solution before electrospinning. For example, Lin et al. [8] added SiO_2 nanoparticles into PS/DMF solution. Figure 2.1 shows the obtained PS fibers with varying content of SiO_2 nanoparticles. Nano-protrusions form at the fiber surfaces and increase with the content of SiO_2 nanoparticles. By adding 7.7 wt% and 14.3 wt% of SiO_2, contact angle increases from 148° to 153° and 157°, and CAH decreases from 7° to 3° and 2°, respectively. The obtained superhydrophobicity is attributed to the nano-protrusions and the grooves on the fiber surfaces, which held liquid from penetrating to the inside of fibrous mat. Based on similar mechanism, TiO_2 nanoparticles and graphene nano-flakes are doped into PS and PCL fibers for superhydrophobicity [7].

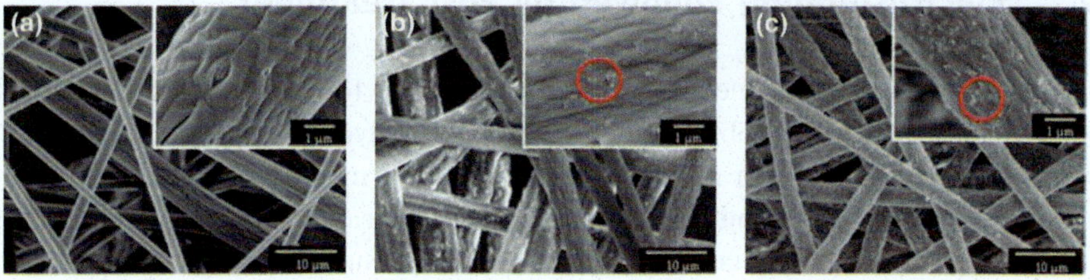

Fig. 2.1 Electrospun PS fibers doped with **a** 0 wt%, **b** 7.7 wt%, and **c** 14.3 wt% of SiO_2 nanoparticles [8]

Without engineering the secondary morphologies of fibers, Wong et al. [9] proposed a method to develop superhydrophobicity by patterning smooth PS fibers. The collector for electrospinning consists of a PDMS substrate and two parallel-arranged electrodes on the top. Before electrospinning, the substrate is pre-stretched with varying strain, i.e., 0–100%. In electrospinning, PS fibers are deposited onto the pre-stretched substrate with alignment by the two parallel-arranged electrodes. After electrospinning, the pre-stretched substrate is relaxed and gets shrunk, resulting in wake-like PS fibers. By increasing the strain of pre-stretched substrate from 0 to 100%, contact angle increases from 138° to 167° and sliding angle decreases from 90° pinning to 5°. The lotus superhydrophobicity can be attributed to the large curvature of the wake-like fibers, which help trap air under droplets and maintain Cassie-Baxter state.

In addition to adjusting fiber morphology, superhydrophobicity may also be reached by lowering fiber surface energy. For one-step method, this can only be realized by adopting low surface energy precursor for electrospinning. However, polymers with very low surface energy like PDMS and poly(tetrafluoroethylene) (PTFE), often have low electrospinnability. To overcome this problem, methods such as polymer blending [10, 13] and coaxial electro-spinning [11] are proposed.

As an example, Ma et al. [10] blended copolymer poly(styrene-*b*-dimethylsiloxane) (PS-PDMS) with PS for electrospinning: PDMS has a very low surface energy of 19.9 mN/m, given the lowest surface energy of 18 mN/m by PTFE [14]; PS has high electrospinnability. The obtained PS-PDMS/PS nanofibers have diameter of 150–400 nm, with a contact angle of 163° and CAH of 15°. The high hydrophobicity is attributed to the aggregation of PDMS blocks at fiber surfaces, as verified by X-ray photoelectron spectrometer. Furthermore, Han et al. [11] produced coaxial fibers with Teflon AF as sheath and PCL as core by coaxial electrospinning. The obtained coaxial fibers have diameter of 1–2 μm with grooves on fiber surface; the coaxial structure is verified by energy-dispersive X-ray spectroscopy. As a result, the contact angle is characterized to be 158° and 10 μL droplets can roll off the surface at a tilting angle of ~7°.

2.2 Electrospinning Followed by Post Treatment

Table 2.2 summarizes the methods of electrospinning followed by post treatment to develop lotus superhydrophobicity. Most of the superhydrophobicity in Table 2.2 are developed in a way that fiber surface roughness is first modified by creating secondary morphology in electrospinning or by post treatment after electrospinning, and then fiber surface energy is reduced by coating a thin layer of low surface energy material. In other words, the superhydrophobicity is achieved by controlling both fiber morphology and fiber surface energy.

In Table 2.2, a few secondary fiber morphologies are included: porous fibers [16], beaded fibers [17–20], and nanoparticle decorated fibers [16, 21–24, 26–28]. Here the nanoparticle decorated fibers are discussed in detail since porous fibers and beaded fibers are introduced in Sects. 1.3 and 2.1. Nanoparticle decorated fibers can be produced by either adding nanoparticles into precursor solution before electrospinning [22–24, 27] or coating nanoparticles onto the fibers after electrospinning [16, 21, 26, 28].

For adding nanoparticles into precursor solution, the nanoparticles may be partially buried in the electrospun fibers, as shown in Fig. 2.2. Post treatment by calcination can roughen fiber surfaces by pyrolysis of polymer matrix. Figure 2.2 compares the morphology of (A) PEO/SiO_2 and (B) PVA/SiO_2 fibers before and after calcination. The calcination exposes the SiO_2 particles by melting or burning out the PEO or PVA, resulting in enhanced fiber surface roughness.

Different from one-step methods, various post treatments allow to coat fibers with nano-particles after electrospinning. Figure 2.3 shows the SEM images of various electrospun fibers coated with SiO_2 nanoparticles by post treatments. In Fig. 2.3a, nylon-6 fibers are coated with positively charged PAH and negatively charged 50 nm SiO_2 nanoparticles by LBL deposition [16]. In Fig. 2.3b, PAN fibers are first modified by ASO using dip coating and then coated by 7–40 nm SiO_2 nanoparticles using blade coating [26]. In Fig. 2.3c, PU fibers are first grafted with -NCO group, and then treated with (3-aminoproyl) triethoxysilane (APTES), and finally coated by SiO_2 particles by hydrothermal method [28]. Compared with the doped fibers in Figs. 2.1 and 2.2, the fibers in Fig. 2.3 show higher surface roughness.

After developing surface structure onto fibers, post treatment such as iCVD, dip coating, physical deposition, can coat fibers with a thin layer of low surface energy material to help maintain droplets in Cassie-Baxter state. As shown in Table 2.2, the low surface energy material can be fluoride such as PPFEMA [16, 17], FDTS [22] and PFDT [24], and silicide such as DMDCS [15]. Note that the coated layer should be thin to reserve the developed surface structure of fibers. For example, Ma et al. [16] fabricated porous PMMA microfibers with average pore size of 80 nm. The porous fibers have a contact angle of 147°. To increase the hydrophobicity of the porous fibers, a PPFEMA layer with thickness of 30 nm is coated. As a result, the PPFEMA coated fibers retained the porous structure and show a contact angle of 163°.

Table 2.2 Electrospinning followed by post treatment for lotus superhydrophobicity

Post treatment methods	Materials	Contact angle	Roll-off angle (droplet volume)	Refs.
CVD	PVDF/DMDCS	153°	–	[15]
iCVD	PMMA/PPFEMA	163°	–	[16]
iCVD	PCL/PPFEMA	175°	2.5° (20 μL)	[17]
Annealing	Poly(AN-co-TMI)/ fluorolink-D/T2EH	167°	4° (5 μL)	[18]
Thermal imidization	PI	160°	–	[19]
CF_4 plasma etching	PVDF	164°	5° (12.5 μL)	[20]
LBL/CVD	Nylon 6/PAH/SiO_2 NPs/ $(CF_3)(CF_2)_5(CH_2)_2SiCl_3$	168°	–	[16]
LBL/FAS modification	Cellulose acetate/PAA/ TiO_2 NPs	162°	2° (12 μL)	[21]
Calcination/ fluorination	PEO/SiO_2 NPs/FDTS PAM/SiO_2 NPs/FDTS PAM/PS/SiO_2 NPs/FDTS	160° 149° 152°	<2° (10 μL)	[22]
Calcination/FAS modification	PVA/silica gel/SiO_2 NPs	155°	3° (12 μL)	[23]
Dip coating	PAN/SiO_2 NPs/Ag/PFDT	157°	–	[24]
Dip coating	SBS/SiO_2 NPs/FDTS	156°	8° (5 μL)	[25]
Dip coating/blade coating	PAN/ASO/SiO_2 NPs	156°	–	[26]
Physical deposition	PU/SiO_2 NPs/DTMS	157°	5° (12 μL)	[27]
Surface grafting/ hydrothermal method	PU/SiO_2 particles	152°	5° (10 μL)	[28]

CVD: chemical vapor deposition; iCVD: initiated chemical vapor deposition; LBL: layer-by-layer deposition; FAS: fluoroalkylsilane

DMDCS: dichlorodimethyl silane; fluorolink-D: perfluorinated linear diol; T2EH: tin(ii) ethyl hexanoate; PI: polyimide; PAH: poly(allylamine hydrochloride); PAA: poly(acrylic acid); PEO: poly(ethylene oxide); FDTS: (heptadecafluoro-1, 1, 2, 2-tetrahydrodecyl)trichlorosilane; PAM: poly(acryl amide); PAN: polyacrylonitrile; PFDT: 1H,1H,2H,2H-perfluorodecanethiol; SBS: poly(styrene-b-butadiene-b-styrene) ASO: amino-silicone oil; DTMS: *n*-dodecyltrimethoxysilane

2.3 Superhydrophobicity with "Rose-Petal" Effect

Superhydrophobicity with "rose-petal" effect, also known as sticky superhydrophobicity, is also important to many applications including microdroplet transportation/collection [29] and lap-on-chip devices [30]. On a sticky superhydrophobic surface, a droplet takes

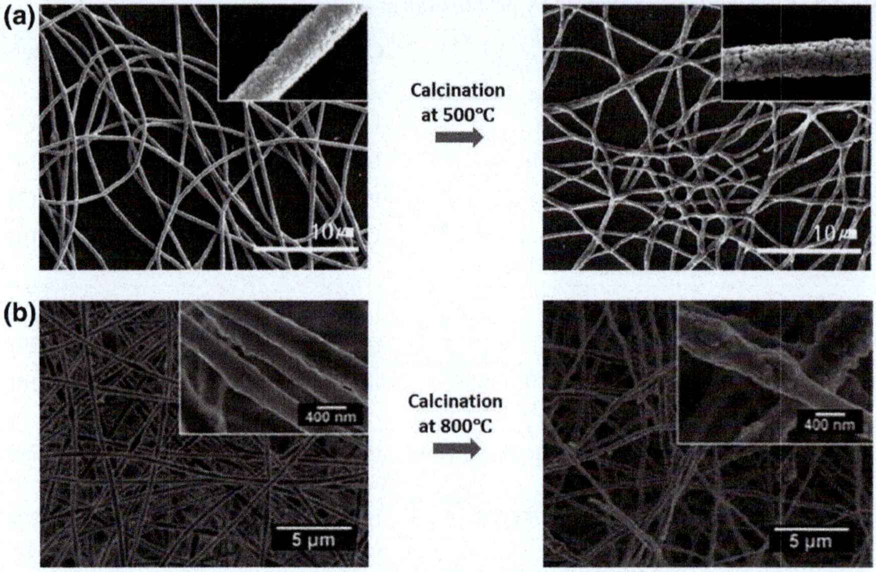

Fig. 2.2 Change of surface morphology of **a** PEO/SiO_2 [22] and **b** PVA/SiO_2 [23] fibers before and after calcination

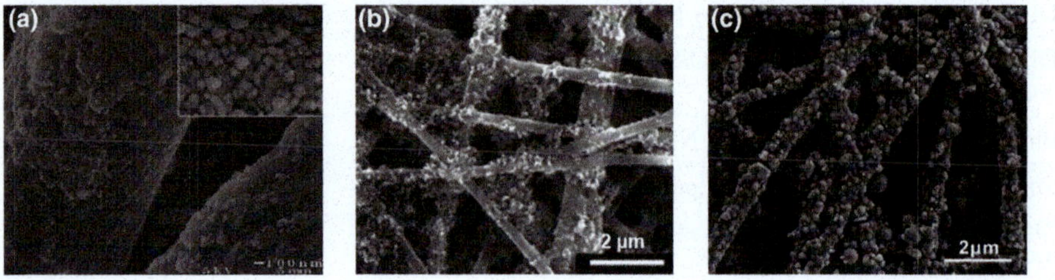

Fig. 2.3 SEM images of **a** nylon-6 fibers [16], **b** PAN fibers [26], and **c** PU fibers [28] coated with SiO_2 nanoparticles after electrospinning

a quasi-spherical shape with contact angle greater than 150°, so the contact area between droplet and solid surface is small. Meanwhile, the droplet can be hold in place without movement because of the high adhesive force by the surface.

Table 2.3 summarizes the electrospinning-based methods for developing rose-petal effect surfaces, which are categorized into one-step electrospinning and electrospinning followed by post treatment. The first column lists the post treatment involved; the second and third columns list the surface morphology and the materials, respectively, of fibers; the fourth and fifth columns show the wetting behaviors on the fibers.

Table 2.3 Electrospinning-based methods for petal superhydrophobic surfaces

Post-treatment	Fiber morphology	Material	Contact angle	Adhesion strength (droplet volume)	Refs.
–	Smooth fibers	CA	154°	90° pinning	[31]
–	Smooth fibers	PolyPhe	160°	180° pinning (2 μL)	[32]
–	Beaded fibers	FPI	158°	98 μN	[33]
–	Smooth fibers	γ-PGA-Phe	154°	180° pinning (1 μL)	[34]
–	Beaded fibers	PVDF	150°	180° pinning (3 μL)	[20]
–	Beaded fibers	PS	152°	113 μN	[35]
Thermal imidization	Beaded fibers	PI	151°	80 μN	[19]
$SiCl_4/H_2O$ treatment/ silanization	Beaded fibers	PVA/silica	168°[a]	180° pinning (2 μL)	[36]

CA: cellulose acetate; PolyPhe: poly(L-phenylalanine); FPI: fluorinated polyimide; γ-PGA-Phe: γ-PGA modified with *L*-phenylalanine ethylester
[a] Advancing contact angle

Lotus fibers require droplets staying in the Cassie-Baxter state, so a proper combination of structured fiber surface and low fiber surface energy is needed. However, for rose-petal fibers, droplets can be in partial wetting or Wenzel state. Therefore, most of the rose-petal fibers in Table 2.3 are produced by one-step electrospinning without post treatment to coat nanoparticles or low surface energy layer. For example, Fig. 2.4 presents three rose-petal fibers that have smooth fiber surface and are fabricated using hydrophilic polymers. Figure 2.4a shows cellulose acetate, which has an intrinsic contact angle of 64°, fibers with diameter of 350–400 nm [31]. As shown in the inset, the cellulose acetate fibers have a contact angle of 154° and droplets are pinned even by tilting the fibrous mat 90°. Figure 2.4b, c are for PolyPhe [32] and γ-PGA-Phe [34] fibers, respectively. The intrinsic contact angle of the PolyPhe and γ-PGA-Phe is 75° and 70°, respectively. Both fibers have contact angle greater than 150° and can pin droplets at a tilting angle of 180°.

Beaded fibers are also widely used to create rose-petal superhydrophobicity. For example, Gong et al. [33] fabricated beaded FPI fibers using one-step electrospinning. Nanopapillae form on the bead surface, increasing the surface roughness of bead. The beaded FPI fibers have contact angle of 158° and adhesion force of 98 μN. The high adhesion is attributed to the noncomposite interface between fibers and droplet, and the hydrophilic groups of FPI molecules. The composite interface between beads and droplet contributes to the high contact angle. In addition, beaded PVDF [20], PS [35], PI [19],

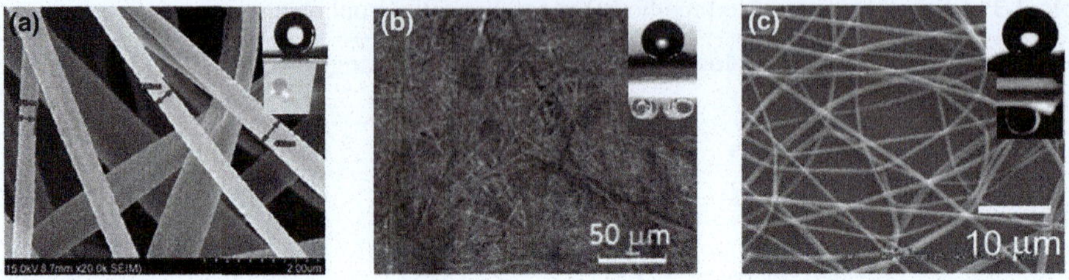

Fig. 2.4 SEM images of superhydrophobic, **a** cellulose acetate [31], **b** PolyPhe [32], and (C) γ-PGA-Phe [34] fibers with rose-petal effect

and PVA/silica [36] fibers are also reported for the feasibility of developing rose-petal superhydrophobicity. Compared to the lotus beaded fibers reported in Tables 2.1 and 2.2, rose-petal beaded fibers have higher fiber surface energy, smaller size contrast between beads and fibers, and lower distribution density of beaded fibers.

2.4 Summary

This chapter first reviews the electrospinning methods to create lotus superhydrophobicity. Based on the fabrication steps, the methods are classified into one-step electrospinning and electrospinning followed by post treatments.

The most commonly used one-step method is creating protrusions on fiber surfaces, i.e., beaded fibers and nanoparticle-decorated fibers. The mechanism for creating lotus effect is that the protrusions can support droplets in Cassie-Baxter state with less continuous liquid-fiber contact interface, resulting in high contact angle and low droplet adhesion. However, the beaded fibers suffer from low mechanical strength due to the high heterogeneity; complex preparation for a homogeneous solution is needed for electrospinning nanoparticle-decorated fibers. Moreover, methods using patterned fibers, coaxial or copolymer electrospinning are also reported, but limited by the complex setup or operation.

For the electrospinning followed by post treatments, most of the lotus-effect surfaces are developed in a procedure that fiber surface roughness is first modified by creating secondary fiber morphology, especially surface protrusions, in electrospinning or by post treatments, and then fiber surface energy is reduced by coating a layer of low surface energy material after electrospinning. Even though the post treatment after electrospinning allows to separate fiber fabrication and surface modification, the fabrication is complicated by the multistep process.

In comparison, developing petal-effect surfaces by electrospinning has less requirement on the fiber morphology and surface energy than that for lotus surfaces. The reason lies in that liquid can penetrate into the inter-fiber spaces so a complete Cassie-Baxter state is not needed for petal effect. Therefore, the petal effect has been achieved using both smooth and beaded fibers with a relatively high surface energy, for example, smooth cellulose acetate nanofibers with random distribution.

Despite the substantial progress in developing electrospun superhydrophobic surfaces, several limitations preventing broad applications exist. First, a strategic approach to superhydrophobicity considering the influences of fiber morphology is needed. Currently, most works rely on increasing fiber surface roughness and/or reducing fiber surface energy with low surface energy materials to realize superhydrophobicity, so fibers are usually randomly deposited with little control. As a result, a thick fibrous mat is usually made to reduce the heterogeneity of fiber distribution, causing a decrease of air permeability and transparency, and also a waste of materials.

Second, current methods often involve complex fabrication processes. To increase fiber surface roughness, nanoparticles are widely adopted before or after electrospinning. Before electrospinning, nanoparticles should be homogeneously blended with the polymer solution, so an extra step might be needed to disperse the nanoparticles. After electrospinning, post treatment is needed to coat fibers with nanoparticles. To reduce the surface energy of fibers, copolymer (low surface energy polymer and electrospinnable polymer) or coaxial electro-spinning, or surface coating after electrospinning are often used. All these methods increase the fabrication complexity.

References

1. L. Jiang, Y. Zhao, J. Zhai, A lotus-leaf-like superhydrophobic surface: a porous microsphere/nanofiber composite film prepared by electrohydrodynamics. Angew. Chem.-Int. Ed. **43**(33), 4338–4341 (2004)
2. Y. Zhu, J. Zhang, Y. Zheng, Z. Huang, L. Feng, L. Jiang, Stable, superhydrophobic, and conductive polyaniline/polystyrene films for corrosive environments. Adv. Func. Mater. **16**(4), 568–574 (2006)
3. M.W. Lee, S. An, S.S. Latthe, C. Lee, S. Hong, S.S. Yoon, Electrospun polystyrene nanofiber membrane with superhydrophobicity and superoleophilicity for selective separation of water and low viscous oil. ACS Appl. Mater. Interfaces **5**(21), 10597–10604 (2013)
4. L. Zhou, C.L. Li, P.T. Chang, S.H. Tan, A.L. Ahmad, S.C. Low, Intrinsic microspheres structure of electrospun nanofibrous membrane with rational superhydrophobicity for desalination via membrane distillation. Desalination **527**, 115594 (2022)
5. V. Pais, M. Navarro, C. Guise, R. Martins, R. Fangueiro, Hydrophobic performance of electrospun fibers functionalized with TiO2 nanoparticles. Text. Res. J. **92**(15–16), 2719–2730 (2022)
6. M. Kang, R. Jung, H.-S. Kim, H.-J. Jin, Preparation of superhydrophobic polystyrene membranes by electrospinning. Colloids Surf. A **313**, 411–414 (2008)
7. R. Asmatulu, M. Ceylan, N. Nuraje, Study of superhydrophobic electrospun nanocomposite fibers for energy systems. Langmuir **27**(2), 504–507 (2011)
8. J. Lin, Y. Cai, X. Wang, B. Ding, J. Yu, M. Wang, Fabrication of biomimetic superhydrophobic surfaces inspired by lotus leaf and silver ragwort leaf. Nanoscale **3**(3), 1258–1262 (2011)

9. W.S. Wong, P. Gutruf, S. Sriram, M. Bhaskaran, Z. Wang, A. Tricoli, Strain engineering of wave-like nanofibers for dynamically switchable adhesive/repulsive surfaces. Adv. Func. Mater. **26**(3), 399–407 (2016)
10. M. Ma, R.M. Hill, J.L. Lowery, S.V. Fridrikh, G.C. Rutledge, Electrospun poly (styrene-block-dimethylsiloxane) block copolymer fibers exhibiting superhydrophobicity. Langmuir **21**(12), 5549–5554 (2005)
11. D. Han, A.J. Steckl, Superhydrophobic and oleophobic fibers by coaxial electrospinning. Langmuir **25**(16), 9454–9462 (2009)
12. Z. Asvar, E. Mirzaei, N. Azarpira, B. Geramizadeh, M. Fadaie, Evaluation of electrospinning parameters on the tensile strength and suture retention strength of polycaprolactone nanofibrous scaffolds through surface response methodology. J. Mech. Behav. Biomed. Mater. **75**, 369–378 (2017)
13. I. Sadeghi, N. Govinna, P. Cebe, A. Asatekin, Superoleophilic, mechanically strong electrospun membranes for fast and efficient gravity-driven oil/water separation. ACS Appl. Polym. Mater. **1**(4), 765–776 (2019)
14. M.E. Callow, R.L. Fletcher, The influence of low surface energy materials on bioadhesion—A review. Int. Biodeterior. Biodegrad. **34**(3–4), 333–348 (1994)
15. F. Zareei Pour, M.M. Sabzehmeidani, H. Karimi, V. Madadi Avargani, M. Ghaedi, Superhydrophobic–superoleophilic electrospun nanofibrous membrane modified by the chemical vapor deposition of dimethyl dichlorosilane for efficient oil–water separation. J. Appl. Polym. Sci. **136**(24), 47621 (2019)
16. M. Ma, M. Gupta, Z. Li, L. Zhai, K.K. Gleason, R.E. Cohen et al., Decorated electrospun fibers exhibiting superhydrophobicity. Adv. Mater. **19**(2), 255–259 (2007)
17. M. Ma, Y. Mao, M. Gupta, K.K. Gleason, G.C. Rutledge, Superhydrophobic fabrics produced by electrospinning and chemical vapor deposition. Macromolecules **38**(23), 9742–9748 (2005)
18. K. Acatay, E. Simsek, C. Ow-Yang, Y.Z. Menceloglu, Tunable, superhydrophobically stable polymeric surfaces by electrospinning. Angew. Chem. **116**(39), 5322–5325 (2004)
19. G. Gong, J. Wu, X. Jin, L. Jiang, Adhesion tuning at superhydrophobic states: from petal effect to lotus effect. Macromol. Mater. Eng. **300**(11), 1057–1062 (2015)
20. H.R. Hong, C.H. Park, The influence of nanostructure on the wetting transition of polyvinylidene fluoride nanoweb: from the petal effect to the lotus effect. Text. Res. J. **91**(7–8), 752–765 (2021)
21. T. Ogawa, B. Ding, Y. Sone, S. Shiratori, Super-hydrophobic surfaces of layer-by-layer structured film-coated electrospun nanofibrous membranes. Nanotechnology **18**(16), 165607 (2007)
22. J.-M. Lim, G.-R. Yi, J.H. Moon, C.-J. Heo, S.-M. Yang, Superhydrophobic films of electrospun fibers with multiple-scale surface morphology. Langmuir **23**(15), 7981–7989 (2007)
23. F. Zhao, X. Wang, B. Ding, J. Lin, J. Hu, Y. Si et al., Nanoparticle decorated fibrous silica membranes exhibiting biomimetic superhydrophobicity and highly flexible properties. RSC Adv. **1**(8), 1482–1488 (2011)
24. T.-T. Li, Y. Wang, H.-K. Peng, X. Zhang, B.-C. Shiu, J.-H. Lin et al., Lightweight, flexible and superhydrophobic composite nanofiber films inspired by nacre for highly electromagnetic interference shielding. Compos. A Appl. Sci. Manuf. **128**, 105685 (2020)
25. D.-E. Lee, E.-Y. Choi, H.-J. Yang, A.S. Murthy, T. Singh, J.-M. Lim et al., Highly stretchable superhydrophobic surface by silica nanoparticle embedded electrospun fibrous mat. J. Colloid Interface Sci. **555**, 532–540 (2019)
26. J. Sheng, Y. Xu, J. Yu, B. Ding, Robust fluorine-free superhydrophobic amino-silicone oil/SiO2 modification of electrospun polyacrylonitrile membranes for waterproof-breathable application. ACS Appl. Mater. Interfaces **9**(17), 15139–15147 (2017)

27. S. Jin, Y. Park, C.H. Park, Preparation of breathable and superhydrophobic polyurethane electrospun webs with silica nanoparticles. Text. Res. J. **86**(17), 1816–1827 (2016)
28. H. Gu, G. Li, P. Li, H. Liu, T.T. Chadyagondo, N. Li et al., Superhydrophobic and breathable SiO2/polyurethane porous membrane for durable water repellent application and oil-water separation. Appl. Surf. Sci. **512**, 144837 (2020)
29. X. Hong, X. Gao, L. Jiang, Application of superhydrophobic surface with high adhesive force in no lost transport of superparamagnetic microdroplet. J. Am. Chem. Soc. **129**(6), 1478–1479 (2007)
30. A.I. Neto, C.R. Correia, C.A. Custódio, J.F. Mano, Biomimetic miniaturized platform able to sustain arrays of liquid droplets for high-throughput combinatorial tests. Adv. Func. Mater. **24**(32), 5096–5103 (2014)
31. F. Mikaeili, P.I. Gouma, Super water-repellent cellulose acetate mats. Sci. Rep. **8**(1), 12472 (2018)
32. H. Yoshida, K. Yanagisawa, Creation of superhydrophobic poly (L-phenylalanine) nonwovens by electrospinning. Polymers **10**(11), 1212 (2018)
33. G. Gong, J. Wu, J. Liu, N. Sun, Y. Zhao, L. Jiang, Bio-inspired adhesive superhydrophobic polyimide mat with high thermal stability. J. Mater. Chem. **22**(17), 8257–8262 (2012)
34. H. Yoshida, D. Klee, M. Möller, M. Akashi, Creation of superhydrophobic electrospun nonwovens fabricated from naturally occurring poly (amino acid) derivatives. Adv. Func. Mater. **24**(40), 6359–6364 (2014)
35. W.S. Wong, N. Nasiri, G. Liu, N. Rumsey-Hill, V.S. Craig, D.R. Nisbet et al., Flexible transparent hierarchical nanomesh for rose petal-like droplet manipulation and lossless transfer. Adv. Mater. Interfaces **2**(9), 1500071 (2015)
36. T. Pisuchpen, N. Chaim-Ngoen, N. Intasanta, P. Supaphol, V.P. Hoven, Tuning hydrophobicity and water adhesion by electrospinning and silanization. Langmuir **27**(7), 3654–3661 (2011)

A Thermodynamic Model for Wetting Various Parallel-Structured Surfaces

3

The wettability of a solid surface is often characterized with a sessile droplet contact angle [1, 2]. A hydrophilic surface has strong affinity to liquid, thereby showing a static contact angle less than 90°. They are needed in anti-fogging glasses [3], heat pipes [4], contact lenses [5], etc. Conversely, a hydrophobic surface, showing a static contact angle greater than 90°, is desirable for self-cleaning [6] and anti-icing [7] surfaces.

Patterning surfaces with parallel structures or grooves is an effective and facile method to control the wettability of solid surfaces. Structures with various cross-section shapes are fabricated, including rectangle and trapezoid structures by photolithography [8], micro-milling [9–11], and laser micromachining [12], and round structures by electrospinning [13–15]. The mechanism of modifying surface wettability by parallel structures is by changing surface roughness.

The wetting state of a liquid droplet on a rough surface can be either composite or noncomposite. In composite state, which is also known as Cassie-Baxter state [16], air pockets form along the solid-liquid interface. In the noncomposite state, or Wenzel state [17], the solid surface under droplet is fully wetted by liquid. The equilibrium contact angle (ECA) on a rough surface can be predicted using Cassie-Baxter model [16] or Wenzel model [17]. In reality, however, the free energy barriers that prevent a wetting system from reaching its global minimum of free energy may deviate the actual contact angle from the ECA [1]. In addition, a free energy barrier may separate composite and noncomposite states and hinder the wetting transition between these two states [18, 19].

Therefore, the energy barriers are important to predict wetting behaviors. Many articles [13, 20–26] have reported thermodynamic models describing the change of free energy in wetting to analyze energy barrier. These models can be divided into two categories based on the calculation of the solid-liquid or liquid-vapor area under droplet. Specifically, some

Y. Zhang, *Superhydrophobicity of Surfaces Dressed by Electrospun Fibers*, Synthesis Lectures on Green Energy and Technology, https://doi.org/10.1007/978-3-031-55552-7_3

works calculate the solid-liquid or liquid-vapor area using a statistical distribution density of structures; the energy barriers in wetting transition for pillars [20–22], spheres [22], and micro hoodoo structures [13] are obtained. However, the energy barriers formed along the movement of exterior three-phase contact line (TPCL) are difficult to analyze with the statistical distribution density.

Alternatively, the solid-liquid or liquid-vapor area can be correlated with the position of exterior TPCL on every structure. For example, Li et al. [23] and He et al. [25] studied the energy barriers in wetting pillar structures with different top-views by calculating the energy difference between discrete positions of exterior TPCL; Tie et al. [24] studied the energy barriers in wetting trapezoid, rectangle, paraboloid, and triangle strips. However, the variation of free energy with the movement of exterior TPCL inside structure grooves was not included. Kim et al. [26] modeled the exterior TPCL moving in rectangle grooves by assuming the liquid front inside grooves as a vertical line, but this assumption cannot fit re-entrant structures, e.g., inverted trapezoid. In addition, numerically calculating the energy difference between discrete TPCL positions can only give the magnitude of energy barriers.

Besides the magnitude, the forming conditions of energy barriers are also important. For example, specifying the forming conditions of energy barrier helps to maximize metastable contact angles for superhydrophobicity [27] and analyze the advancing and receding contact angles for contact angle hysteresis [23]. To this end, Shuttleworth and Bailey [28] studied the movement of exterior TPCL on wavy surfaces (with inclination angle less than 90°) by considering intrinsic contact angle (ICA) as local contact angle. They modeled the initiating and ending conditions of the energy barrier in noncomposite state. Later, Eick et al. [27] related free energy to apparent contact angle (APCA) as droplets spread along sawtooth surfaces. They derived similar equations by minimizing system free energy. More recently, Long et al. [29] extended the application of the model for the initiating and ending conditions of energy barrier to parallel-structured surfaces. Moreover, other scholars model contact angle hysteresis, which is same with the initiating and ending conditions of energy barriers without external perturbations [26], for pillars [30, 31], spheres [32], and re-entrant structures [33] in composite state. However, the forming conditions of energy barrier may be discontinuous and complicated by (1) the interplay between ICA and the multiple surface inclination angles of round or quadrilateral structures, and (2) the penetration of exterior TPCL inside the grooves [14].

In addition, the reported thermodynamic models [10, 14, 24, 26] relating free energy with the exterior TPCL position for parallel structures are shape specific. There is a lack of an integrated model for various simple geometries, such as trapezoid [11, 12], rectangle [8–10], inverted trapezoid [34], circle [13, 14], and elongated circle [15]. Besides evaluating the wetting behaviors of existing structures, an integrated model can also guide the design of new structures.

Therefore, the objectives of this chapter are to build an integrated thermodynamic model relating free energy with the continuous movement of TPCLs and then specify

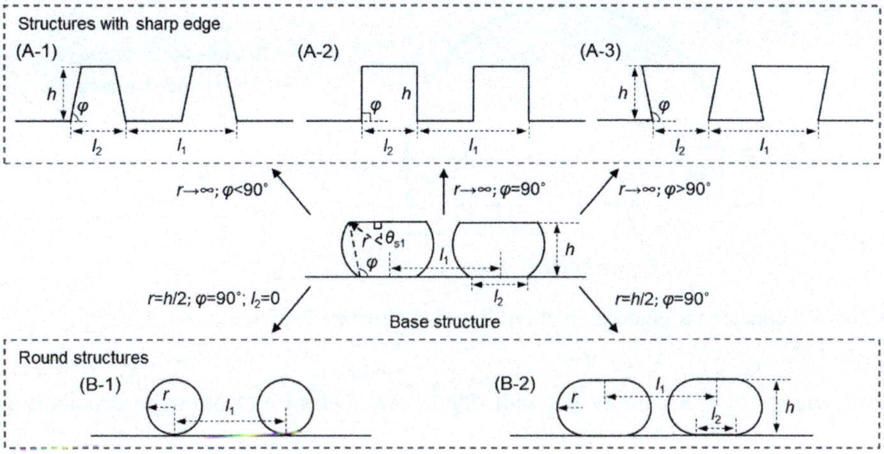

Fig. 3.1 Cross-section view of various structures: (A-1) trapezoid; (A-2) rectangle; (A-3) inverted trapezoid; (B-1) circle; (B-2) elongated circle

the forming conditions of energy barrier in wetting various parallel-structured surfaces. Figure 3.1 shows the cross section of the studied parallel structures: (A) sharp-edge and (B) round structures, which are often reported in the literature. As illustrated, both types of structures are specific case of the base structure. This analysis enables the development of an integrated model in this work. Accordingly, the rest of the chapter is organized as follows. Section 3.1 introduces the thermodynamic models for different wetting states. The critical conditions for creating energy barriers are derived from the models. Then Sect. 3.2 validated the thermodynamic models and the forming conditions of energy barrier using experimental data by electrospun fibers and existing experimental and theoretical works. Section 3.3 compares the forming conditions and magnitude of energy barrier for various parallel structures.

3.1 Theoretical Analyses

Figure 3.2 presents the schematic of a liquid droplet on a parallel-structured surface. Two-dimensional thermodynamic models relating system free energy with the position of TPCLs are built for the composite and noncomposite states, and wetting transition. The assumptions are made as follows: (1) the liquid-vapor interface 1 is treated as a circular segment with a radius of R_1 by ignoring the effect of gravity for droplet smaller than capillary length [35]; (2) the liquid-vapor interface 2 is assumed planar by ignoring the effect of Laplace pressure [35]; (3) the line tension is ignored so Young equation is locally valid [36]; (4) the droplet volume remains constant by ignoring droplet evaporation.

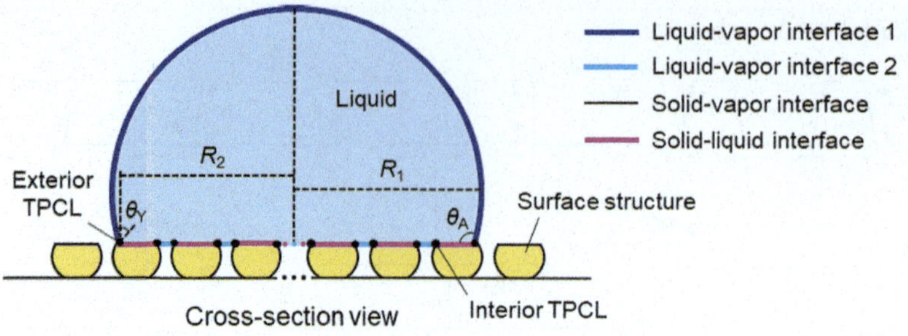

Fig. 3.2 Schematic of a liquid droplet on a parallel-structured surface

The change of free energy per unit depth, ΔE, before and after the droplet contacts the structure surface in Fig. 3.2 is:

$$\Delta E = (L_{LV1} + L_{LV2})\gamma_{LV} + L_{SL}(\sigma_{SL} - \sigma_{SV}) \tag{3.1}$$

where L and γ are interface length and surface tension, respectively; L, V, and S represent liquid, vapor, and solid, respectively. Then dividing Eq. (3.1) by γ_{LV} and applying the Young equation [37, 38] give

$$\Delta E/\gamma_{LV} = L_{LV1} + L_{LV2} - L_{SL}\cos\theta_Y \tag{3.2}$$

where θ_{Y} is the ICA of solid surface. For a given volume of a liquid, the system free energy before contacting and γ_{LV} are constant; thus, the $\Delta E/\gamma_{\text{LV}}$ of Eq. (3.2) stands for system free energy per unit depth for the rest of this chapter. Based on the geometric relationship shown in Fig. 3.2 and the constant cross-section area, S, of the droplet [24, 39], we have

$$L_{LV1} = 2R_1\theta_A = 2R_2\theta_A/\sin\theta_A \tag{3.3}$$

$$S = (\theta_A - \sin\theta_A\cos\theta_A)R_2^2/\sin^2\theta_A + S_2 \tag{3.4}$$

where θ_{A} is the APCA, and S_2 is the area difference between the liquid below external TPCL and the structure above external TPCL to account for the liquid penetration into the grooves. The L_{LV2} and L_{SL} in Eq. (3.2) are functions of APCA and the position of TPCLs, and they depend on the wetting states, which are specified as follows.

Figure 3.3 shows the composite and noncomposite states, and wetting transition for the sharp-edge and round structures in Fig. 3.1. Different from earlier 2D models [23, 24], the model in this paper considers the movements of exterior TPCL along each structure, so the energy barrier encountered at each surface can be analyzed. As the droplet spreads along the structures, the exterior TPCL moves in a cyclic pattern. The position of the exterior TPCL is described using n, s, δ_{o} and θ_{o}: $2(n-1)$ is the number of structure units that

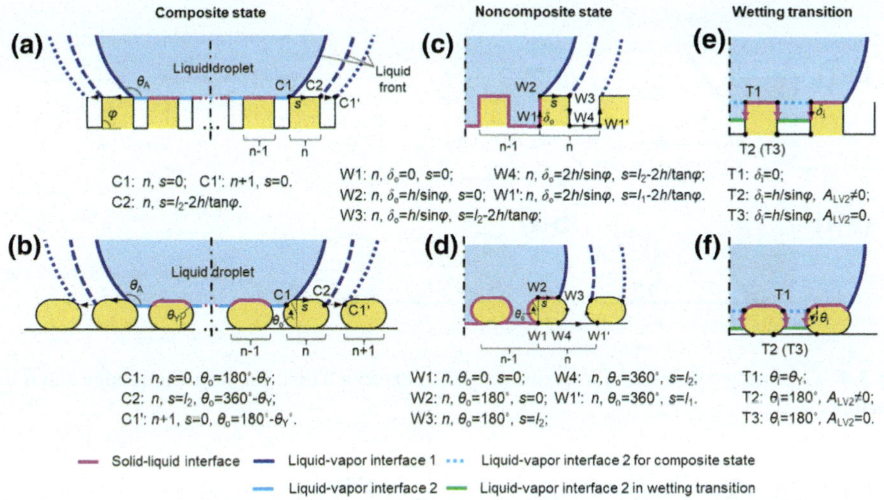

Fig. 3.3 Schematics of composite states for the **a** sharp-edge and **b** round structures; noncomposite states for the **c** sharp-edge and **d** round structures; the wetting transition for **e** sharp-edge and **b** round structures. (trapezoid: $\varphi < 90°$; rectangle: $\varphi = 90°$; inverted trapezoid: $\varphi > 90°$; circle: $l_2 = 0$; elongated circle: $l_2 \neq 0$)

have been fully covered by droplet; s and δ_o are the wetted length of the nth structure unit along horizontal and inclined surfaces, respectively; θ_o is the angular position of exterior TPCL. The interior TPCL uses δ_i or θ_i for inclined or curved surfaces.

However, a few geometry constraints limit the movement of exterior TPCL, because the liquid front cannot intercept with structures at any point besides exterior TPCL. Figure 3.4a, b illustrate the geometry constraints for the exterior TPCL on the sharp-edge and round structures, respectively. As shown in Fig. 3.4(A-1), the two green dotted lines set the boundary of the liquid front, which is denoted by the blue line. It should be noted that the liquid front inside grooves is treated as a straight line, which is tangent to the liquid front curve outside grooves [14, 29], for simplicity. As a result, there are upper and lower limits for θ_A corresponding to boundary of the liquid front.

For the sharp-edge structures in Fig. 3.4a, there is no constraint for the path 2–3, i.e., when the external TPCL is on path 2–3, because 2–3 is the top surface. For the paths 3–4, 4–1', and 1'–2', which correspond to Fig. 3.4(A-1)–(A-3), respectively, the constraints, i.e., the upper and lower limits of θ_A, are expressed using Eq. (3.5), which is derived based on geometric relationships.

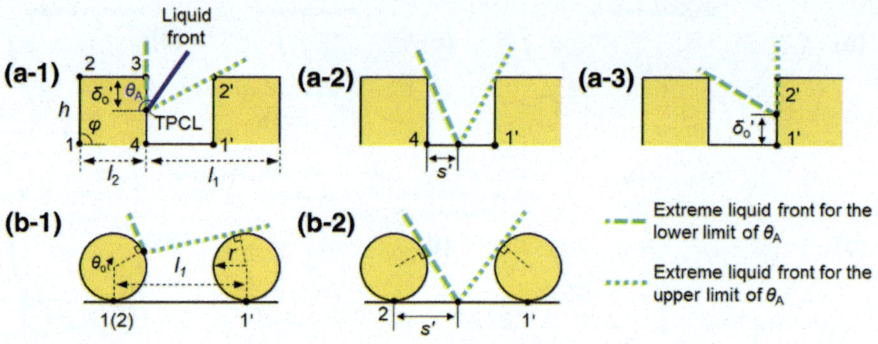

Fig. 3.4 Geometry constraints for the movement of exterior TPCL on the **a** sharp-edge and **b** round structures

$$
\begin{aligned}
&3-4{:}\varphi < \theta_A < 180^\circ - \arctan\frac{\delta_0'\sin\varphi}{l_1 - l_2 + 2h/\tan\varphi - \delta_0\cos\varphi}\\
&4-1'{:}\arctan\frac{h}{h/\tan\varphi + s'} < \theta_A < 180^\circ - \arctan\frac{s'\sin\varphi}{l_1 - l_2 + h/\tan\varphi - s'}\\
&1'-2'{:}\arctan\frac{h-\delta_0'\sin\varphi}{l_1 - l_2 + h/\tan\varphi + \delta_0'\cos\varphi} < \theta_A < 180^\circ - \varphi
\end{aligned}
\tag{3.5}
$$

where φ, l_1, l_2, and h are all geometry parameters as introduced in Fig. 3.1; s' and δ_o' are the wetted length of each horizontal and non-horizontal surfaces, respectively, by exterior TPCL. For the round structures,

$$
\begin{aligned}
&1-2: \text{arc}\sin\frac{r}{\sqrt{(l_1 - r\sin\theta_0)^2 + (r\cos\theta_0)^2}}\\
&+\text{arc}\ \tan\frac{r\cos\theta_0}{l_1 - r\sin\theta_0} < \theta_A < \theta_0 \text{ if } \theta_0 < 180^\circ\\
&\theta_0 - 180^\circ < \theta_A < 180^\circ - \text{arc}\sin\frac{r}{\sqrt{(l_1 - r\sin\theta_0)^2 + (r\cos\theta_0)^2}}\\
&-\text{arc}\tan\frac{r\cos\theta_0}{l_1 - r\sin\theta_0} \text{ if } \theta_0 > 180^\circ\\
&2-1': 2\text{arc}\tan\frac{r}{s'} < \theta_A < 180^\circ - 2\text{arc}\tan\frac{r}{l_1 - s'}
\end{aligned}
\tag{3.6}
$$

where r and l_1 are geometry parameters; θ_0 is the location of external TPCL on each round structure.

3.1.1 Geometric Transformation

Figure 3.5 defines the geometry of the base structure in Fig. 3.1. The positions of exterior and interior TPCLs are described using θ_o and θ_i in Fig. 3.5a, b, respectively. In both figures, the gold (color) region represents a single base structure, and its side walls are two arcs with a radius of r. As r approaches infinity, the arc walls become straight, so the base structure becomes a sharp-edge structure with an internal angle of φ, as shown Fig. 3.1a. With a finite r, the base structure becomes an elongated circle when $h = 2r$ and $\varphi = 90°$, and then becomes a circle as l_2 equals zero, as shown Fig. 3.1b.

In the geometric transformation from the base to sharp-edge structures, as r approaches infinity, the exterior TPCL at the point C in Fig. 3.5a moves right to the point D, and the interior TPCL at the point C' in Fig. 3.5b moves right to the point D'. So, the heights of exterior TPCL (length of KJ) and interior TPCL (length of K'J) are fixed at $\delta_o' \cdot \sin\varphi$ and $h - \delta_i \cdot \sin\varphi$, respectively. Then Eqs. (3.7–3.8) are obtained based on the geometric relationship of $KJ = OE \cdot \cos(\angle JOE) - OC \cdot \cos\theta_o$ and $LK = OB \cdot \cos(\angle BOK') - OC' \cdot \cos\theta_i$, respectively.

$$\delta_o' \sin\varphi = r\cos(180° - 2\varphi + \theta_{s1}) - r\cos\theta_o \tag{3.7}$$

$$\delta_i \sin\varphi = r\cos\theta_{s1} - r\cos\theta_i \tag{3.8}$$

where r, φ, and $\theta_{s1} = \varphi - \arcsin(h/(2r\sin\varphi))$ are all structure parameters. Meanwhile, for the exterior TPCL, the length of arc CE approaches δ_o' and the difference between the projections of OC and OE on CD approaches $-\delta_o' \cdot \cos\varphi$. Then the following relations can be obtained,

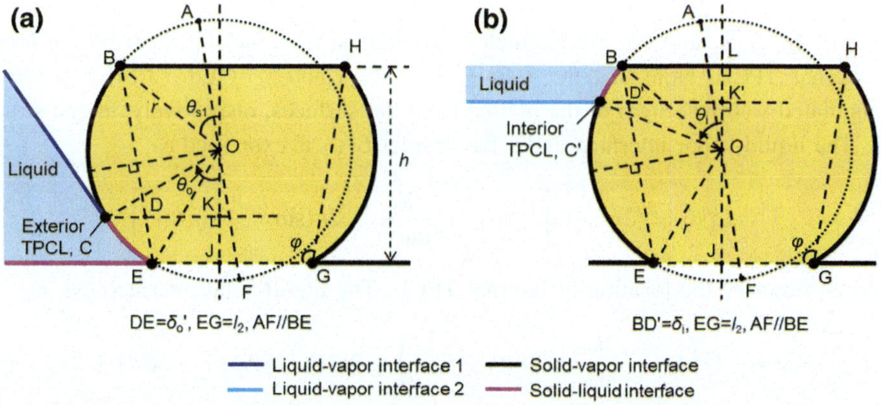

Fig. 3.5 Geometry of the base structure in Fig. 3.1 with **a** exterior and **b** interior TPCLs

$$\lim_{r\to\infty} r(\theta_o - 180° + \theta_{s2}) = \delta_o' \tag{3.9}$$

$$\lim_{r\to\infty} r\sin\theta_o - r\sin\theta_{s2} = -\delta_o'\cos\varphi \tag{3.10}$$

where θ_{s2} is defined as $2\varphi - \theta_{s1}$ for $\theta_o < 180°$, and $-\theta_{s1}$ for $\theta_o > 180°$ to simplify Eqs. (3.9) and (3.10). Similarly, for the interior TPCL in Fig. 3.5b,

$$\lim_{r\to\infty} r(\theta_i - \theta_{s1}) = \delta_i \tag{3.11}$$

$$\lim_{r\to\infty} r\sin\theta_i - r\sin\theta_{s1} = \delta_i\cos\varphi \tag{3.12}$$

The geometric relationships of Eqs. (3.7)–(3.12) are used to integrate various structures.

3.1.2 Composite State

In the composite state, the exterior TPCL moves in a cyclic pattern of C1–C2–C1' as shown in Fig. 3.3a, b. Substituting Eq. (3.3) into Eq. (3.2) gives,

$$\frac{\Delta E}{\gamma_{LV}} = 2R_2\frac{\theta_A}{\sin\theta_A} + L_{LV2} - L_{SL}\cos\theta_Y \tag{3.13}$$

where R_2, L_{LV2}, and L_{SL} are specified based on the geometric relationships in Figs. 3.3a, b and 3.5. For the base structure, the half basal width, R_2, is expressed as Eq. (3.14),

$$R_2 = \frac{l_1 - l_2}{2} + (n-1)l_1 - r(\sin\theta_o - \sin\theta_{s2}) + s \tag{3.14}$$

where φ, l_1, l_2, h, and θ_{s2} are geometry parameters; n, s, and θ_o describe the location of external TPCL. Note that the $r(\sin\theta_0 - \sin\theta_{s2})$ and s of Eq. (3.14) represent the accumulated wetting length of the inclined and flat surfaces, respectively, in one structure unit. The liquid-vapor interface under the droplet L_{LV2} is expressed as

$$L_{LV2} = (2n-1)\left[l_1 - l_2 + \frac{2h}{\tan\varphi} - 2r(\sin\theta_i - \sin\theta_{s1})\right] \tag{3.15}$$

where θ_i describe the location of internal TPCL. The liquid-solid interface L_{SL} is

$$L_{SL} = 2(n-1)\left(l_2 - \frac{2h}{\tan\varphi} + 2r(\theta_i - \theta_{s1})\right) + 2r(\theta_o - (\theta_o)_{C1} - 2\theta_{s3}) + 2s \tag{3.16}$$

where $(\theta_o)_{C1}$ is the angle of initial point in the cyclic pattern of C1–C2–C1'; $\theta_{s3} = 0$ for $\theta_o < 180°$ and $\theta_{s3} = \theta_{s1}$ for $\theta_o > 180°$. Equations (3.13)–(3.16) are the integrated model for composite state.

Particularly, for the sharp-edge structures in Fig. 3.1, $\delta_i = 0$; $\delta_o' = (\delta_o')_{C1} = h/\sin\varphi$ for $\theta_o < 180°$ and $\delta_o' = 0$ for $\theta_o > 180°$; $\theta_o = (\theta_o)_{C1} + 2\theta_{s3}$. The integrated model can be specified into the sharp-edge model by substituting Eqs. (3.10), (3.12), and (3.11) into Eqs. (3.14–3.16), respectively.

$$\begin{aligned}\frac{\Delta E}{\gamma_{LV}} &= \left[\left(l_1 - l_2 + \frac{2h}{\tan\varphi}\right) + 2(n-1)l_1 + 2s\right]\frac{\theta_A}{\sin\theta_A} \\ &+ (2n-1)\left(l_1 - l_2 + \frac{2h}{\tan\varphi}\right) - \left[2(n-1)\left(l_2 - \frac{2h}{\tan\varphi}\right) + 2s\right]\cos\theta_Y \end{aligned} \tag{3.17}$$

For the round structures, $\theta_i = 180° - \theta_Y$ and $(\theta_o)_{C1} = \theta_Y$. The model is specified into

$$\begin{aligned}\frac{\Delta E}{\gamma_{LV}} &= [l_1 - l_2 + 2(n-1)l_1 - 2r\sin\theta_o + 2s]\frac{\theta_A}{\sin\theta_A} + (2n-1)(l_1 - l_2 - 2r\sin\theta_Y) \\ &- \left[2(n-1)\left(l_2 + 2r\left(180° - \theta_Y\right)\right) + 2r(\theta_o - \theta_Y) + 2s\right]\cos\theta_Y \end{aligned} \tag{3.18}$$

The next step is to analyze the variation of free energy with the exterior TPCL. The $\Delta E/\gamma_{LV}$ varies continuously as the exterior TPCL moves from C1 to C2. Then the variation of $\Delta E/\gamma_{LV}$ with s for the sharp-edge structures is calculated by taking the derivative of Eq. (3.17) with respect to s,

$$\frac{d(\Delta E/\gamma_{LV})}{ds} = 2R_2\frac{d(\theta_A/\sin\theta_A)}{dR_2}\frac{dR_2}{ds} + 2\left(\frac{\theta_A}{\sin\theta_A} - \cos\theta_Y\right) \tag{3.19}$$

From the derivative of Eq. (3.4), where S_2 is zero in composite state, with respect to R_2,

$$\frac{d\theta_A}{dR_2} = \frac{\sin\theta_A(\theta_A - \sin\theta_A\cos\theta_A)}{R_2(\theta_A\cos\theta_A - \sin\theta_A)} \tag{3.20}$$

Substituting Eq. (3.20) into Eq. (3.19) gives,

$$\frac{d(\Delta E/\gamma_{LV})}{ds} = 2(\cos\theta_A - \cos\theta_Y) \tag{3.21}$$

Then the variation of $\Delta E/\gamma_{LV}$ with s from C1 to C2 can be determined as,

$$\begin{cases} \dfrac{d(\Delta E/\gamma_{LV})}{ds} < 0\,\theta_A > \theta_Y \\ \dfrac{d(\Delta E/\gamma_{LV})}{ds} > 0\,\theta_A < \theta_Y \end{cases} \tag{3.22}$$

Based on the same method, Eq. (3.23) can be obtained for C1–C2 of the round structures,

$$\begin{cases} \frac{d(\Delta E/\gamma_{LV})}{d\theta_o} < 0 \, \theta_A - (180^\circ - \theta_Y) < \theta_o < \theta_A + (180^\circ - \theta_Y) \\ \frac{d(\Delta E/\gamma_{LV})}{d\theta_o} > 0 \, \theta_o < \theta_A - (180^\circ - \theta_Y) \; or \; \theta_o > \theta_A + (180^\circ - \theta_Y) \end{cases} \tag{3.23}$$

For either type of structure, the exterior TPCL jumps from C2 to C1' because there is no intermediate point in between. From Eq. (3.17), the change of $\Delta E/\gamma_{LV}$ from C2 to C1' is expressed as

$$\Delta\left(\frac{\Delta E}{\gamma_{LV}}\right)_{C2-C1'} = 2\left(l_1 - l_2 + \frac{2h}{\tan\varphi}\right) + \left(\frac{2R_2\theta_A}{\sin\theta_A}\right)_{C1'} - \left(\frac{2R_2\theta_A}{\sin\theta_A}\right)_{C2} \tag{3.24}$$

The derivative of $2R_2\theta_A/\sin\theta_A$ with respect to R_2 is

$$\frac{d(2R_2\theta_A/\sin\theta_A)}{dR_2} = 2\frac{\theta_A}{\sin\theta_A} + 2R_2\frac{\sin\theta_A - \theta_A\cos\theta_A}{\sin^2\theta_A}\frac{d\theta_A}{dR_2} \tag{3.25}$$

Substituting Eq. (3.20) into Eq. (3.25) gives

$$\frac{d(2R_2\theta_A/\sin\theta_A)}{dR_2} = 2\cos\theta_A \tag{3.26}$$

Then Eq. (3.24) can be re-written as

$$\Delta\left(\frac{\Delta E}{\gamma_{LV}}\right)_{C2-C1'} = 2\left(\left(l_1 - l_2 + \frac{2h}{\tan\varphi}\right) + \int_{R_2}^{R_2+\left(l_1-l_2+\frac{2h}{\tan\varphi}\right)} \cos\theta_A dR_2\right) \tag{3.27}$$

where $\cos\theta_A$ is higher than -1, so the integration is always larger than $-(l_1-l_2 + 2h/\tan\varphi)$. As a result, $\Delta(E/\gamma_{LV})_{C2-C1'}$ is always positive, indicating that free energy increases from C2 to C1'.

Figure 3.6 illustrates the general approach to determine the forming conditions of energy barrier. After calculating the critical APCAs, the full APCA range, [0°, 180°], is divided into several sub-ranges by the critical APCAs. The next step is to determine the local minima of free energy, which indicate the formation of energy barrier, by analyzing the variation of $\Delta E/\gamma_{LV}$ along one movement cycle of external TPCL for each APCA sub-range. Then the APCA sub-ranges that form energy barrier are located and the forming conditions of energy barrier are specified. Take the sharp-edge structures in composite state as an example. The critical APCA of θ_Y from Eq. (3.22) divides the full APCA range into [0°, θ_Y] and [θ_Y, 180°]. In [0°, θ_Y], $\Delta E/\gamma_{LV}$ keeps decreasing as the

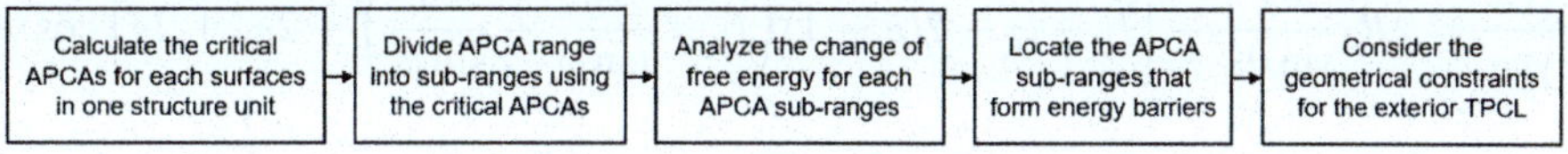

Fig. 3.6 Procedure to specify the forming conditions of energy barriers

external TPCL moves from C1–C2–C1'. In [θ_Y, 180°], $\Delta E/\gamma_{LV}$ first decreases along C1–C2 and then increases along C2–C1', creating a local minima at C2 and an energy barrier forms. For a higher accuracy, unrealistic energy barriers due to the geometry constraints (see Fig. 3.4) for TPCL should be removed.

In the composite state, the forming conditions of energy barrier are as follows. For the sharp-edge structures, energy barrier forms when APCA is greater than θ_Y, i.e., [θ_Y, 180°]. The round structures can extend the range of APCA for energy barrier to [$2\theta_Y - 180°$, 180°] because the surface of a round structure has continuous inclination angles to the horizontal plane. However, the APCA range for round structures may shrink because of the geometry constraints, which can only be determined with known structure parameters and thereby presented as part of the results in Sect. 3.3.

3.1.3 Noncomposite State

Figure 3.3c, d present the noncomposite state. In this case, exterior TPCL can enter the grooves and move in the cyclic pattern of W1–W2–W3–W4–W1' or W1–W2–W1' for circle. Mathematically, $L_{LV2} = 0$ for no liquid-vapor interface under droplet, resulting in

$$\frac{\Delta E}{\gamma_{LV}} = 2R_2\frac{\theta_A}{\sin\theta_A} - L_{SL}\cos\theta_Y \tag{3.28}$$

where R_2 and L_{SL} are specified based on the geometric relationships in Figs. 3.3c, d and 3.5. The half basal width R_2 is the same with that for composite state, as expressed in Eq. (3.14). The liquid-solid interface L_{SL} is

$$L_{SL} = l_1 - l_2 + 2(n-1)\left[l_1 - \frac{2h}{\tan\varphi} + 2r(\theta_i - \theta_{s1})\right] + 2r(\theta_o - (\theta_o)_{W1} - 2\theta_{s3}) + 2s \tag{3.29}$$

where $(\theta_o)_{W1}$ is the angle of initial point in the noncomposite cyclic pattern; $\theta_{s3} = 0$ for $\theta_o < 180°$ and $\theta_{s3} = \theta_{s1}$ for $\theta_o > 180°$. Equations (3.28), (3.14) and (3.29) are the integrated model for noncomposite state. Particularly, $\delta_i = h/\sin\varphi$ and $(\delta_e)_{W1} = 0$ for the sharp-edge structures. The integrated model becomes the sharp-edge model by substituting Eqs. (3.10) and (3.11) into Eqs. (3.14) and (3.29), respectively.

$$\frac{\Delta E}{\gamma_{LV}} = 2R_2\frac{\theta_A}{\sin\theta_A} - \left[l_1 - l_2 + 2(n-1)\left(l_1 - \frac{2h}{\tan\varphi} + \frac{2h}{\sin\varphi}\right) + 2\delta_o + 2s\right]\cos\theta_Y \tag{3.30}$$

For the round structures, $\theta_i = 180°$ and $(\theta_o)_{W1} = 0°$. The model is specified into

$$\frac{\Delta E}{\gamma_{LV}} = 2R_2\frac{\theta_A}{\sin\theta_A} - \left[l_1 - l_2 + 2(n-1)\left(l_1 - \frac{2h}{\tan\varphi} + 2\pi r\right) + 2r\theta_o + 2s\right]\cos\theta_Y \tag{3.31}$$

For both the sharp-edge and round structures, the variations of $\Delta E/\gamma_{LV}$ with s along W2–W3 and W4–W1$'$ (or W2–W1$'$ for circle) are the same as that for C1–C2. For W2–W3 and W4–W1', Eq. (3.4) becomes

$$\begin{aligned} &\text{W2} - \text{W3}: \ R_0^2\pi = R_2^2\frac{\theta_A - \sin\theta_A\cos\theta_A}{\sin^2\theta_A} + (2n-1)S_{groove} \\ &\text{W4} - \text{W1}': \ R_0^2\pi = R_2^2\frac{\theta_A - \sin\theta_A\cos\theta_A}{\sin^2\theta_A} - 2nS_{structure} \end{aligned} \tag{3.32}$$

where S_{groove} is the cross-section area of single groove between two adjacent structures, and $S_{structure}$ is the cross-section area of one structure. However, S_{groove} and $S_{structure}$ are constants when TPCL moves from W2 to W3 or W4–W1', respectively. Thus, Eqs. (3.20) and (3.22) are still valid.

For W3–W4 of the sharp-edge structures, $\Delta E/\gamma_{LV}$ with δ_o, so Eq. (3.30) is specified as,

$$\begin{aligned} \frac{\Delta E}{\gamma_{LV}} &= 2R_2\frac{\theta_A}{\sin\theta_A} \\ &-\left[l_1 - l_2 + 2(n-1)\left(l_1 - \frac{2h}{\tan\varphi} + \frac{2h}{\sin\varphi}\right) + 2\left(l_2 - \frac{2h}{\tan\varphi} + \frac{h}{\sin\varphi}\right) + 2\delta_o\right]\cos\theta_Y \end{aligned} \tag{3.33}$$

The derivative of Eq. (3.33) with respect to δ_o is

$$\frac{d(\Delta E/\gamma_{LV})}{d\delta_o} = 2R_2\frac{\sin\theta_A - \theta_A\cos\theta_A}{\sin^2\theta_A}\frac{d\theta_A}{d\delta_o} + 2\left(\cos\varphi\frac{\theta_A}{\sin\theta_A} - \cos\theta_Y\right) \tag{3.34}$$

The Eq. (3.4) becomes

$$R_0^2\pi = R_2^2\frac{\theta_A - \sin\theta_A\cos\theta_A}{\sin^2\theta_A} + (2n-1)S_{groove'} - 2nS_{structure'} \tag{3.35}$$

where $S_{groove'}$ and $S_{structure'}$ are the partial area of groove and structure, respectively, as shown in Fig. 3.7. From Fig. 3.7, the following relationship can be obtained,

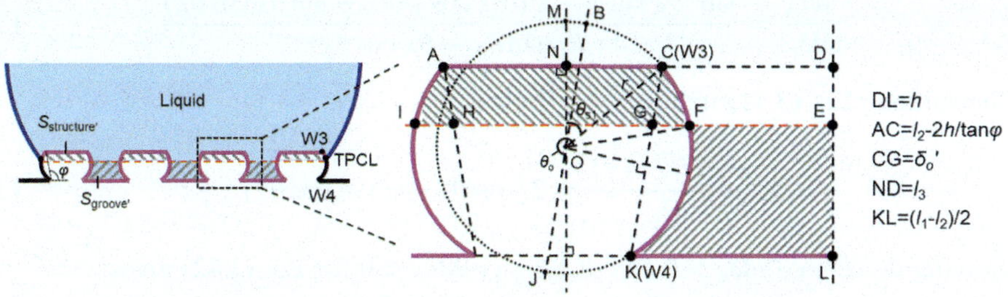

Fig. 3.7 Schematics of $S_{\mathrm{groove}'}$ and $S_{\mathrm{structure}'}$ for the base structure

$$S_{\mathrm{structure}'} = S_{\mathrm{AHGC}} + 2S_{\mathrm{CGF}} \tag{3.36}$$

where S_{AHGC} and S_{CGF} are expressed as Eqs. (3.37) and (3.38), respectively, based on the geometric relationships shown in the inset of Fig. 3.7.

$$S_{\mathrm{AHGC}} = \left(l_2 - \frac{2h}{\tan\varphi} + \delta_o' \cos\varphi\right)\delta_o' \sin\varphi \tag{3.37}$$

$$\begin{aligned} S_{\mathrm{CGF}} = &\frac{r^2}{2}\left[\theta_{\mathrm{o}} - \pi - \theta_{\mathrm{s1}} + \frac{1}{2}\sin(2\theta_{\mathrm{s1}}) - \frac{1}{2}\sin(2\theta_{\mathrm{o}})\right] \\ &-\left(r\sin\theta_{\mathrm{s1}} + \frac{1}{2}\delta_o'\cos\varphi\right)\delta_o'\sin\varphi \end{aligned} \tag{3.38}$$

For $S_{\mathrm{groove}'}$, point K can be treated as the location of the interior TPCL. Then $S_{\mathrm{groove}'}$ can be calculated by

$$\begin{aligned} S_{\mathrm{groove}'} = &2S_{\mathrm{EFKL}} = 2l_3(\delta_{\mathrm{i}} - \delta_o')\sin\varphi \\ &-r^2\left[\theta_{\mathrm{o}} - \pi - \theta_{\mathrm{s1}} - \theta_{\mathrm{i}} - \frac{1}{2}\sin\theta_{\mathrm{o}} + \frac{1}{2}\sin(2\theta_{\mathrm{i}} + 2\theta_{\mathrm{s1}})\right] \end{aligned} \tag{3.39}$$

where l_3 is expressed by $l_1/2 - l_2/2 + h/\tan\varphi + r\sin\theta_{\mathrm{s1}}$. In composite state, $\theta_{\mathrm{i}} = \pi - \theta_{\mathrm{Y}}$ or 0 when there is no penetration of droplet. In noncomposite state, $\theta_{\mathrm{i}} = 2(\varphi - \theta_{\mathrm{s1}})$. In wetting transition, θ_{i} varies with penetration depth. For the sharp-edge structures in noncomposite state, $S_{\mathrm{groove}'}$ and $S_{\mathrm{structure}'}$ are specified as

$$\begin{aligned} S_{\mathrm{structure}'} &= \delta_o'\left(l_2 - \frac{2h}{\tan\varphi} + \delta_o'/\tan\varphi\right) \\ S_{\mathrm{groove}'} &= (h - \delta_o')\left(l_1 - l_2 + (h - \delta_o')/\tan\varphi\right) \end{aligned} \tag{3.40}$$

After substituting Eq. (3.40) into Eqs. (3.35) and (3.41) is obtained from the derivative of Eq. (3.35) with respect to δ_o' (same with δ_{o} for derivative).

$$\frac{d\theta_A}{d\delta_o} = \frac{\sin^3\theta_A \sin\varphi - \sin\theta_A \cos\varphi(\theta_A - \sin\theta_A \cos\theta_A)}{R_2(\sin\theta_A - \theta_A \cos\theta_A)} \tag{3.41}$$

Substituting Eq. (3.41) into Eq. (3.34) gives

$$\frac{d(\Delta E/\gamma_{LV})}{d\delta_o} = 2(\cos(\theta_A - \varphi) - \cos\theta_Y) \tag{3.42}$$

Solving $d(\Delta E/\gamma_{LV})/d\delta_o > 0$ and $d(\Delta E/\gamma_{LV})/d\delta_o < 0$ for Eq. (3.42) gives

$$\begin{cases} \frac{d(\Delta E/\gamma_{LV})}{d\delta_o} > 0 \varphi - \theta_Y < \theta_A < \varphi + \theta_Y \\ \frac{d(\Delta E/\gamma_{LV})}{d\delta_o} < 0 \theta_A < \varphi - \theta_Y \; or \; \varphi + \theta_Y < \theta_A \end{cases} \tag{3.43}$$

For W1–W2, Eq. (3.44) can be obtained by the same derivation method with that of W3–W4.

$$\frac{d(\Delta E/\gamma_{LV})}{d\delta_o} = 2(\cos(\theta_A + \varphi) - \cos\theta_Y) \tag{3.44}$$

Solving $d(\Delta E/\gamma_{LV})/d\delta_o > 0$ and $d(\Delta E/\gamma_{LV})/d\delta_o < 0$ for Eq. (3.44) gives

$$\begin{cases} \frac{d(\Delta E/\gamma_{LV})}{d\delta_o} < 0 \theta_Y - \varphi < \theta_A < 360^\circ - (\varphi + \theta_Y) \\ \frac{d(\Delta E/\gamma_{LV})}{d\delta_o} > 0 \theta_A < \theta_Y - \varphi \text{ or } 360^\circ - (\varphi + \theta_Y) < \theta_A \end{cases} \tag{3.45}$$

For the round structures, the critical APCAs are the same as those shown in Eq. (3.23). However, the restriction of $\theta_Y < \theta_o < 360^\circ - \theta_Y$ in composite state is unneeded in noncomposite state. Thus, a local minimum of $\Delta E/\gamma_{LV}$ at $\theta_o = \theta_A + (180^\circ - \theta_Y)$ can be attained for each round structure.

The forming conditions of energy barrier in noncomposite state depend on the ICA and structure parameters because the relative size of the multiple critical APCAs is needed for the procedure in Fig. 3.6. Therefore, a case study is shown in Sect. 3.3 to illustrate the use of this model.

3.1.4 Wetting Transition

Figure 3.3e, f shows the wetting transition from composite to noncomposite state: liquid penetrates into the surface grooves, and finally the liquid-vapor interface under the droplet disappears. The level of penetration is described by the position of the interior TPCL, i.e., the penetration length (δ_i) and angle (θ_i) for the inclined and curved surfaces, respectively. Additionally, the exterior TPCLs are fixed at C2 in the wetting transition.

Considering that the exterior TPCLs are fixed at C2, the free energy per unit depth for the base structure at any penetration level between T1 and T2 is calculated by Eq. (3.46), which is derived from Eqs. (3.14)–(3.16).

$$\frac{\Delta E}{\gamma_{LV}} = 2R_2\frac{\theta_A}{\sin\theta_A} + (2n-1)\left(l_1 - l_2 + \frac{2h}{\tan\varphi} - 2r(\sin\theta_i - \sin\theta_{s1})\right)$$
$$-\left[2n\left(l_2 - \frac{2h}{\tan\varphi}\right) + 2(2n-1)r(\theta_i - \theta_{s1}) + 2r(\theta_o - (\theta_o)_{C1} - 2\theta_{s3})\right]\cos\theta_Y \tag{3.46}$$

where $r(\theta_o - (\theta_o)_{C1} - \theta_{s3})$ is constant. Particularly, $(\delta_i)_{C1} = 0$ for the sharp-edge structures and $(\theta_i)_{C1} = 180° - \theta_Y$ for the round structures. Then the variation of free energy per unit depth is analyzed by differentiating $\Delta E/\gamma_{LV}$ with respect to θ_i. Solving $d(\Delta E/\gamma_{LV})/d\theta_i = 0$ leads to

$$\cos\theta_i + \cos\theta_Y = -\frac{\sin\theta_A}{2R_2}\left[l_1 - l_2 + \frac{2h}{\tan\varphi} - 2r(\sin\theta_i - \sin\theta_{s1})\right]\sin\theta_i \tag{3.47}$$

In the case of neglecting the influence of liquid penetration (into grooves) on θ_A [36], θ_A can be treated constant. Then the right side of Eq. (3.47), from the $d(2R_2\theta_A/\sin\theta_A)/d\theta_i$, becomes zero. Then for the sharp-edge structures, $\Delta E/\gamma_{LV}$ increases from T1 to T2 when $\theta_Y > 180° - \varphi$, leading to an energy barrier. For the round structures, $\Delta E/\gamma_{LV}$ always increases from T1 to T2 because $180° - \theta_Y < \theta_i < 180°$, so an energy barrier forms too.

From T2 to T3, the liquid-vapor interface 2 disappears ($L_{LV2} = 0$), so the change of free energy per unit depth is

$$\Delta\left(\frac{\Delta E}{\gamma_{LV}}\right)_{T2-T3} = -(2n-1)(l_1 - l_2)(1 + \cos\theta_Y) \tag{3.48}$$

where $(\Delta E/\gamma_{LV})_{T3}$ is calculated by Eq. (3.28). The Eq. (3.48) is always negative, indicating that T2–T3 process is thermodynamically spontaneous. Conversely, $\Delta(E/\gamma_{LV})_{T3\text{–}T2}$ is the free energy barrier to be overcome for the transition from noncomposite to composite states.

Then we evaluate the stability at T1, which corresponds to the composite state, by the difference between $(\Delta E/\gamma_{LV})_{T1}$ from Eq. (3.46) and $(\Delta E/\gamma_{LV})_{T3}$:

$$\Delta\left(\frac{\Delta E}{\gamma_{LV}}\right)_{T1-T3} = -(2n-1)\left(l_1 - l_2 + \frac{2h}{\tan\varphi} - 2r(\sin(\theta_i)_{T1} - \sin\theta_{s1})\right)$$
$$+ 2R_2\left[\left(\frac{\theta_A}{\sin\theta_A}\right)_{T3} - \left(\frac{\theta_A}{\sin\theta_A}\right)_{T1}\right]$$
$$+ (2n-1)\left[2r((\theta_i)_{T1} - (\theta_i)_{T3}) - (l_1 - l_2)\right]\cos\theta_Y \tag{3.49}$$

where a positive $\Delta(E/\gamma_{LV})_{T1-T3}$ indicates a stable composite state, while a negative $\Delta(E/\gamma_{LV})_{T1-T3}$ indicates either a metastable or unstable composite state. Particularly, for the sharp-edge structures, $(\delta_i)_{T1} = 0$ and $(\delta_i)_{T3} = h/\sin\varphi$. Then neglecting the change of θ_A and equating Eq. (3.49) to zero give

$$\cos\theta_Y = -\frac{l_1 - l_2 + 2h\cot\varphi}{l_1 - l_2 + 2h\csc\varphi} \tag{3.50}$$

For the round structures, $(\theta_i)_{T1} = 180° - \theta_Y$ and $(\theta_i)_{T3} = 180°$. Then Eq. (3.51) is obtained.

$$\cos\theta_Y = -\frac{l_1 - l_2 + 2r\sin\theta_Y}{l_1 - l_2 + 2r\theta_Y} \tag{3.51}$$

3.2 Model Validation

The preceding models are validated using data in the literature and those experimentally obtained in our lab. First, the thermodynamic model is validated with experimental data in reference. Then the forming conditions of energy barriers in noncomposite state are verified with an existing model and validated by experimental data using electrospun microfibers. Last, the model for wetting transition is validated with an existing model and experimental data from the literature.

Figure 3.8 compares our thermodynamic model for composite state with the experiment data and theoretical models from Ding et al. [9] for rectangle structures and Wang et al. [12] for trapezoid structures in terms of the equilibrium contact angle (ECA). Specifically, our model relates free energy to APCA, so ECA can be predicted by the global minimization of free energy. Our model accurately interprets the experiment data with mean differences of 1.85% for rectangles (vs. 3.49% in Ref. [9]) and 1.89% for trapezoids (vs. 3.06% in Ref. [12]), respectively. Thus, our thermodynamic model is validated and can be used to analyze free energy barrier.

Figure 3.9 shows the verification of our model for the forming conditions of free energy barrier with the model from Eick et al. [27]. The forming conditions of energy barrier for sharp-edge structures in the noncomposite state are specified using Eqs. (3.22), (3.44), and (3.45) and following the procedure in Fig. 3.6. Here we use hydrophobic materials (i.e., $120° > \theta_Y > 90°$) as an example. Three critical APCAs from the Eqs. (3.22), (3.44), and (3.45) divide the full APCA range of [0°, 180°] into four sub-ranges. The variation of free energy in each sub-range is analyzed for the cyclic pattern of W1–W2–W3–W4–W1′ to determine the formation of energy barriers. As a result, the solid lines with arrows denote our model for sharp-edge structures, covering triangle, trapezoid, rectangle, and inverted trapezoid as φ increases from $\arctan(2h/l_2)$ to 90° or greater. The line with stars shows the model for triangle reported by Eick et al. [27], and it coincides with our model

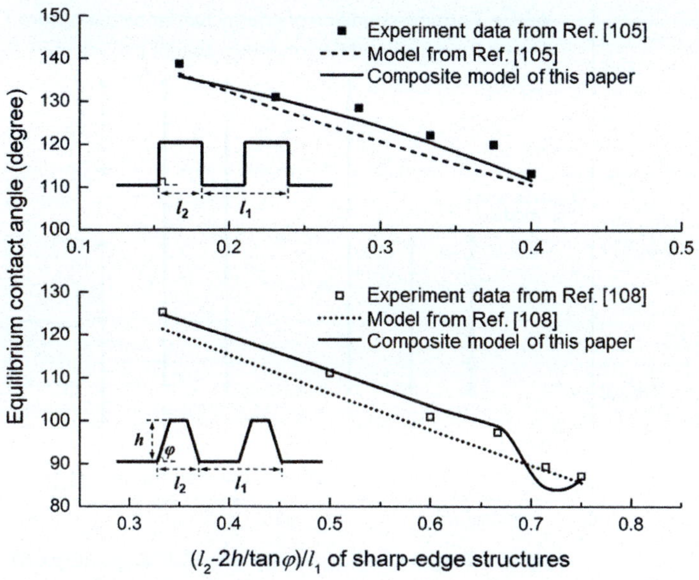

Fig. 3.8 Validation of the thermodynamic model with experimental data in the literature [9, 12]

for the triangle case. Thus, our model is partially verified, and our model can extend to quadrilateral structures. It should be noted that the forming conditions of energy barriers are discontinuous when $\theta_Y > 180° - \varphi/2$ because of the monotonical variation of free energy for $\theta_Y > \theta_A > 180° - |180° - \theta_Y - \varphi|$.

For the round strucures, model or experimental data for the forming conditions of energy barrier in the noncomposite state are rarely found in the literture. Thus, we experimentally validate our model using aligned electrospun microfibers. These fibers can be treated as the round structures shown in Fig. 3.1(B-1). The details for fabrication and characterization of the fibers can be found in Sect. 4.1. Figure 3.10 shows the validation with experiment data. Based on Eq. (3.23), a free energy barrier forms at each round structure if the exterior TPCL is not limited by the geometry constraints. This statement is validated by that three aligned fibers (with a mean radius of 0.9 μm) are able to pin the droplet from spreading and no energy barrier forms at the middle fiber because of the droplet deformation by gravity. Then the APCA is predicted by Eqs. (3.4) and (3.31) at the first advancing energy barrier. Meanwhile, the ECA calculated using Eqs. (3.4) and (3.31) is shown for comparison. The prediction fits with the experimental data with a mean difference of 2.30%.

Figure 3.11 verifies our model for the wetting transition using the model form Vrancken et al. [40]. The critical ICAs separating stable composite and noncomposite states for the sharp-edge structures are calculated using Eq. (3.50) and plotted as the dashed lines, which exemplify trapezoid, rectangle, and inverted trapezoid when φ

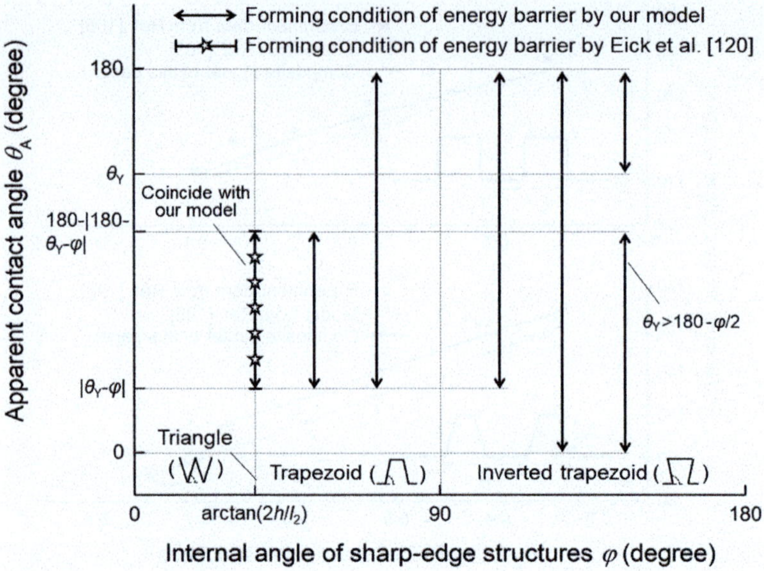

Fig. 3.9 Verification of the forming condition of energy barriers with model reported in reference; the model in Ref. [27] is for triangle structure, a specific case of our model at $\varphi = \arctan(2\ h/l_2)$

Fig. 3.10 Validation of the forming condition of energy barrier for round structure using aligned electrospun microfibers ($r = 0.9\ \mu$m)

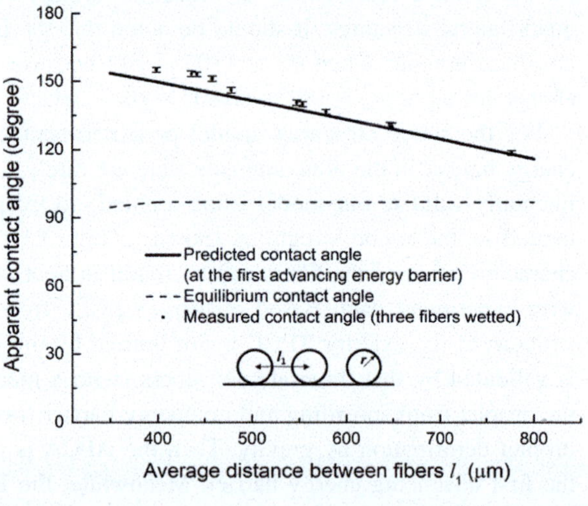

increases from 80° to 110°. Here the change of APCA due to droplet penetration into grooves is neglected. As depicted by the inset, the region above or under the critical ICA line indicates that the composite or noncomposite state is thermodynamically stable. The solid line is the modeled critical ICA line for the rectangle structure reported by Vrancken et al. [40]; it coincides with our model when φ is 90°. In addition, the square points in

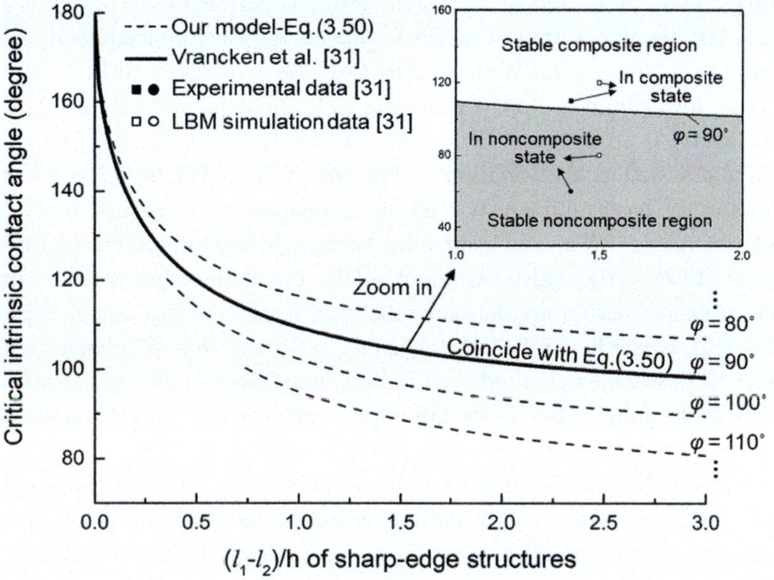

Fig. 3.11 Validation of the model for evaluating the stability of composite wetting state with the results in reference; the model in Ref. [40] is for rectangle structure, a specific case of this paper at $\varphi = 90°$

the inset are predicted to be in composite state, which agrees with the observations in the experiment and Lattice Boltzmann Method (LBM) simulation. Then the ICAs of square points are reduced to below critical ICA line, as depicted by the round hollow points, via electrowetting. As a result, wetting transition is observed in experiment and LBM simulation [40].

3.3 Results and Discussion

This section compares the energy barriers of the structures in Fig. 3.1 using the validated models with specific structure parameters and ICA. The structures adopted have the same height and spacing, and our focus is on comparing the shapes of structures: $l_1 = 10\ \mu m$, $l_2 = 4\ \mu m$, $h = 3\ \mu m$, $r = 1.5\ \mu m$, $\theta_Y = 95°$. The internal angle φ of the sharp-edge structures varies from 65° to 115°, and the droplet volume is 2 μL.

Figure 3.12 specifies the forming conditions of energy barrier in the noncomposite state following the procedure in Fig. 3.6 without considering the geometry constraints. The critical APCAs for energy barrier are calculated using Eqs. (3.22), (3.43) and (3.45) for the sharp-edge structures and Eq. (3.23) for round structures. For the trapezoid with $\varphi = 65°$, energy barrier forms at W3, where the external TPCL locates, when APCA is between θ_Y and $\theta_Y + \varphi$, and at W2 when APCA is between $\theta_Y - \varphi$ and θ_Y. When φ increases to 90°, the upper limit of APCA for the energy barrier increases to 180° because the critical APCA, $\theta_Y + \varphi$ (at W3), exceeds 180°. As φ increases to 115° for an inverted trapezoid, the lower limit of APCA decreases to 0° since the critical APCA, $\theta_Y - \varphi$ (at W2), is below 0°.

For circular structure, the variation of free energy from W1 to W2 is curved because the round surface has continuously varying inclination angle against horizontal plane; energy barrier may form at every structure because a local minimum of free energy at $\theta_o = \theta_A + (180° - \theta_Y)$ exists within W1–W2. For the elongated circle, the forming conditions of energy barrier are the same: the local minima of free energy are at $\theta_o = \theta_A + (180° - \theta_Y)$, which lies in W3–W4 when $\theta_A > \theta_Y$ and W1–W2 when $\theta_A < \theta_Y$.

Figure 3.13 shows the magnitude and forming conditions of energy barrier considering the geometry constraints. For the sharp-edge structures, the APCA ranges for energy barrier are independent on the geometry constraints: $[\theta_Y, 180°]$ for composite state and $[\theta_Y - \varphi, \theta_Y + \varphi]$ for noncomposite state. In composite state, droplet sits on the top surfaces (C1–C2) of structures, so the movement of the external TPCL is not limited by the geometry constraints. In noncomposite state, energy barrier forms at W3 and W2 when APCA is in the range of $[\theta_Y, \theta_Y + \varphi]$ and $[\theta_Y - \varphi, \theta_Y]$, respectively. Even though

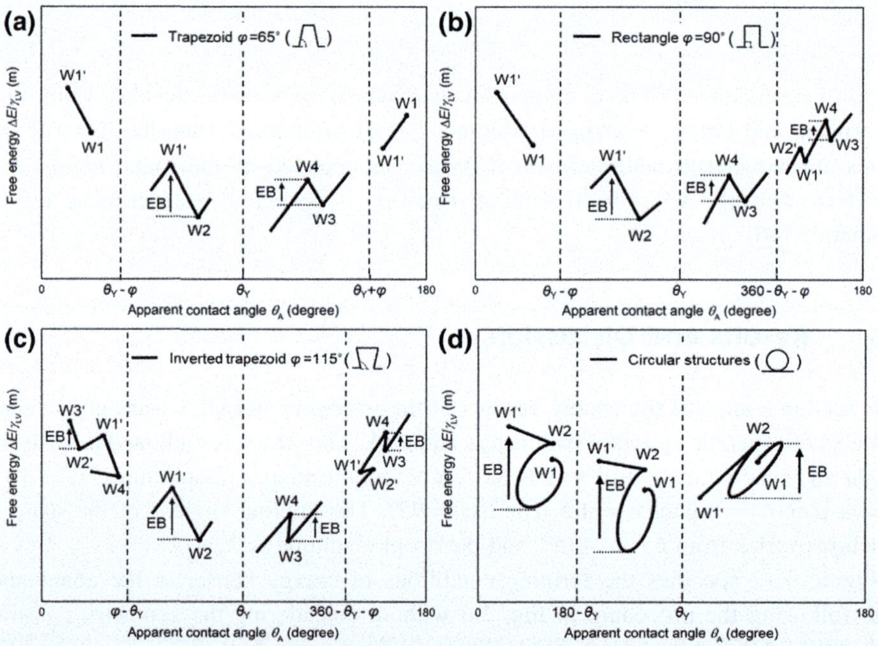

Fig. 3.12 The forming conditions of energy barrier for **a** trapezoid, **b** rectangle, **c** inverted trapezoid, and **d** the round structures in noncomposite state ($\theta_Y = 95°$; EB: energy barrier)

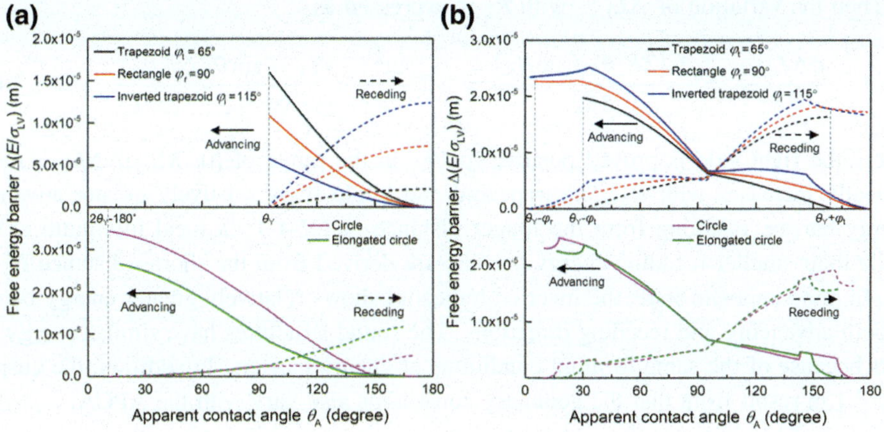

Fig. 3.13 The magnitudes of energy barrier in the **a** composite and **b** noncomposite states

the exterior TPCL is limited by the geometry constraints, it can still encounter the energy barrier at W3 for large APCA, i.e., around $\theta_Y + \varphi$, as shown in Fig. 3.13b, and at W2 for small APCA, i.e., $\theta_Y - \varphi$. However, the APCA range may be discontinuous as shown in Fig. 3.9: for hydrophobic materials, an ICA more than $180° - \varphi/2$ causes a monotonical variation of free energy when APCA is between $[360° - \theta_Y - \varphi, \theta_Y]$ based on Eqs. (3.22), (3.43) and (3.45) and Fig. 3.6. Similarly, for hydrophilic materials, an ICA below $\varphi/2$ results in a monotonical variation of free energy when APCA lies between $[\theta_Y, \varphi - \theta_Y]$. For round structures, the APCA range decreases to [10°, 171°] for circle and [17°, 165°] for elongated circle. The reduction is due to the short gap distance between adjacent structures, preventing the exterior TPCL from reaching the local minimum at $\theta_o = \theta_A + (180° - \theta_Y)$.

In summary, the round structures extend the range of APCA, especially the lower limit, that forms energy barrier in composite state. This indicates that round structures may cause higher contact angle hysteresis along the direction perpendicular to the surface structure in developing superhydrophobic surfaces. In noncomposite state, the re-entrant structures, i.e., inverted trapezoid and round structures, can increase the APCA range for energy barrier. This result can guide the development of "rose petal" effect for applications such as water harvesting [41] and droplet transportation [42].

In Fig. 3.13, both advancing (solid line) and receding (dash line) energy barriers are presented. In composite state, the energy barriers increase toward the moving direction of liquid front because of the monotonically increased ΔL_{LV1}. Based on Eq. (3.2), L_{LV1} and L_{LV2} dominate the change of free energy per unit depth, so the magnitude of energy barrier is mostly influenced by ΔL_{LV1} since ΔL_{LV2} is constant. From Eq. (3.26),

$$\Delta L_{LV1} = \frac{d(2R_2\theta_A/\sin\theta_A)}{dR_2} = 2\cos\theta_A \tag{3.52}$$

Then the variation of ΔL_{LV1} with R_2 is expressed as

$$\frac{\mathrm{d}\Delta L_{\mathrm{LV1}}}{\mathrm{d}R_2} = \frac{\mathrm{d}^2(2R_2\theta_{\mathrm{A}}/\sin\theta_{\mathrm{A}})}{\mathrm{d}R_2^2} = -2\frac{\sin^2\theta_{\mathrm{A}}}{R_2}\frac{(\theta_{\mathrm{A}} - \sin\theta_{\mathrm{A}}\cos\theta_{\mathrm{A}})}{(\theta_{\mathrm{A}}\cos\theta_{\mathrm{A}} - \sin\theta_{\mathrm{A}})} \tag{3.53}$$

where the right side is proved positive for θ_{A} in the range of [0, π], so ΔL_{LV1} monotonically increases with R_2. The trapezoid and circle show relatively greater advancing energy barrier, resulting from the longer distance of C2–C1′ (a local minimum at C2), while their smaller receding energy barriers are derived from the shorter distance of C1–C2. In noncomposite state, the inverted trapezoid shows relatively greater energy barriers in both advancing and receding directions. The round structures have similar energy barriers because of the same forming conditions of energy barriers. In addition, the cusps in Fig. 3.13b result from that the geometry constraints also vary with the APCA.

Figure 3.14a presents the variation of free energy per unit depth with the penetration depth in wetting transition given that the APCAs equal to the ECAs in the composite state, which are calculated to be 153°, 129°, 113°, 138°, and 112° for trapezoid, rectangle, inverted trapezoid, circle, and elongated circle, respectively, using Eqs. (3.17) and (3.18). As the internal TPCLs move from T1 to T2, an energy barrier forms for the rectangle, inverted trapezoid, and round structures because free energy increases according to Eq. (3.47). Moreover, the energy barrier for the re-entrant structures, i.e., inverted trapezoid and round structures, is greater because of the increased liquid-vapor interface under the droplet with penetration depth. From T2 to T3, the liquid-vapor interface under the droplet suddenly disappears, causing a reduction of free energy for all structures. Conversely, the wetting transition from noncomposite to composite needs to overcome this energy difference between T2 and T3.

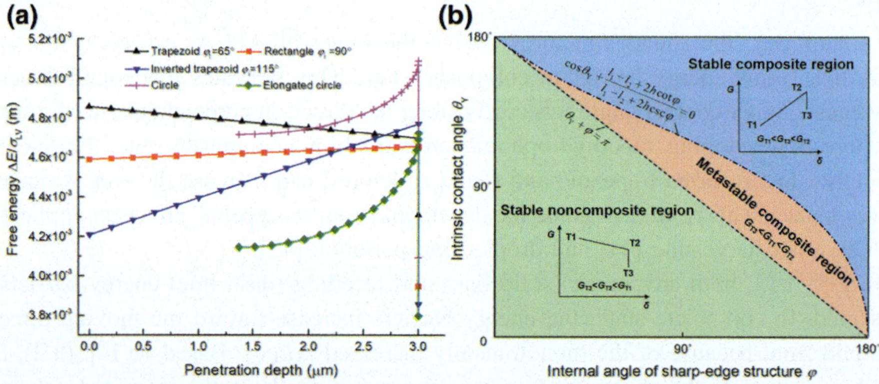

Fig. 3.14 **a** Variation of free energy per unit depth in wetting transition at the ECAs in the composite state; **b** the stability of wetting states for the sharp-edge structures with varying internal angle and ICA ($l_1 = 10$ μm, $l_2 = 4$ μm, $h = 3$ μm)

Figure 3.14b presents the stability of wetting states for the sharp-edge structures. The two critical (dashed) lines are plotted by substituting structure parameters into Eqs. (3.47) and (3.50) by ignoring the change of APCA in wetting transition. Depending on the relative sizes of free energy per unit depth at T1, T2, and T3, the whole map is divided into stable noncomposite, metastable and stable composite regions. In the stable noncomposite region, free energy decreases from T1 to T3, so noncomposite state is energetically favored. In the metastable composite region, noncomposite state is more stable than composite state but the wetting transition from composite to noncomposite state needs to overcome an energy barrier. In the stable composite region, composite state is thermodynamically more stable. Practically, the area of the stable composite region is discounted because a material with θ_Y above 120° is rarely found [43]. For the round structures, two regions, i.e., metastable and stable composite regions, exist based on Eqs. (3.47) and (3.51) by ignoring the sagging effect. In both regions, an energy barrier forms in transition to noncomposite state for the increased free energy from T1 to T2. In reverse, there is also an energy barrier for transition from noncomposite to composite state due to the increased free energy from T3 to T2.

3.4 Summary

This chapter reports an integrated thermodynamic model, which relates system free energy to the continuous movement of three-phase contact lines, for five parallel structures based on geometric transformation. Earlier works [27, 28] reported the initiating and ending conditions of free energy barriers for inclined surfaces. In this work, the forming conditions of energy barriers for the sharp-edge and round structures are derived from the integrated model, and validated with experimental data from our lab and existing model. The specific results and implications are as follows.

(1) For sharp-edge structures with an internal angle φ and intrinsic contact angle θ_Y, the initiating and ending APCAs for energy barriers in noncomposite state are $\theta_Y + \varphi$ and $\theta_Y - \varphi$, respectively. Therefore, increasing φ can expand the APCA range for forming energy barrier. However, the APCA range may be discontinuous when θ_Y is smaller than $\varphi/2$ or larger than $180° - \varphi/2$.

(2) For round structures in noncomposite state, an energy barrier may form for every structure at $\theta_A + (180° - \theta_Y)$ because the round surface has continuous inclination angles with respect to the horizontal plane. However, the geometry constraints can decrease the APCA range for energy barriers to [10°, 171°] for the structure parameters of this work. Increasing the gap or reducing the height of the structure can relax the geometry constraints.

(3) In practice, round structures, e.g., aligned electrospun microfibers (with a mean radius of 0.9 μm), are able to pin the droplet at the first advancing energy barrier. Metastable contact angles greater than 150° are feasible. Therefore, surfaces patterned with round structures can achieve large contact angle with high contact angle hysteresis. These findings are helpful in developing rose petal effect for applications like microdroplet transportation and water harvesting.

Overall, this work provides insights into the interfacial behavior, especially in metastable state, of wetting rough surfaces. The integrated model can guide the design of structures to control surface wettability. However, the relationship between metastable contact angles and the magnitude of energy barrier is unknown. Therefore, quantifying the influences of energy barrier on metastable contact angle is needed in future work. Moreover, analyzing the energy barriers for hierarchically roughened fibers deserves attention in the future.

References

1. C.H. Kung, P.K. Sow, B. Zahiri, W. Mérida, Assessment and interpretation of surface wettability based on sessile droplet contact angle measurement: challenges and opportunities. Adv. Mater. Interfaces **6**(18), 1900839 (2019)
2. N. Janardan, M.V. Panchagnula, Effect of the initial conditions on the onset of motion in sessile drops on tilted plates. Colloids Surf. A **456**, 238–245 (2014)
3. L. Yang, X. Luo, W. Chang, Y. Tian, Z. Wang, J. Gao et al., Manufacturing of anti-fogging super-hydrophilic microstructures on glass by nanosecond laser. J. Manuf. Process. **59**, 557–565 (2020)
4. T. Hao, X. Ma, Z. Lan, N. Li, Y. Zhao, H. Ma, Effects of hydrophilic surface on heat transfer performance and oscillating motion for an oscillating heat pipe. Int. J. Heat Mass Transf. **72**, 50–65 (2014)
5. T. Shimizu, T. Goda, N. Minoura, M. Takai, K. Ishihara, Super-hydrophilic silicone hydrogels with interpenetrating poly (2-methacryloyloxyethyl phosphorylcholine) networks. Biomaterials **31**(12), 3274–3280 (2010)
6. R. Blossey, Self-cleaning surfaces—Virtual realities. Nat. Mater. **2**(5), 301–306 (2003)
7. T.V. Charpentier, A. Neville, P. Millner, R.W. Hewson, A. Morina, Development of anti-icing materials by chemical tailoring of hydrophobic textured metallic surfaces. J. Colloid Interface Sci. **394**, 539–544 (2013)
8. Y. Chen, B. He, J. Lee, N.A. Patankar, Anisotropy in the wetting of rough surfaces. J. Colloid Interface Sci. **281**(2), 458–464 (2005)
9. W. Ding, Z. Cao, B. Wang, S. Xu, Z. Wang, Construction of grating structure model based on Gibbs free energy and experimental verification by micro-milling. Appl. Phys. A **125**(6), 1–8 (2019)
10. S. Zhenyu, L. Zhanqiang, S. Hao, Z. Xianzhi, Prediction of contact angle for hydrophobic surface fabricated with micro-machining based on minimum Gibbs free energy. Appl. Surf. Sci. **364**, 597–603 (2016)
11. C.T. Cheng, S. To, G. Zhang, Characterization of intermediate wetting states on micro-grooves by water droplet contact line. J. Ind. Eng. Chem. **91**, 69–78 (2020)

12. B. Wang, Y. Zhang, J. Song, Z. Wang, Investigation and prediction on regulation of hydrophobicity of polymethyl methacrylate (PMMA) surface by femtosecond laser irradiation. Coatings **10**(4), 386 (2020)
13. A. Tuteja, W. Choi, J.M. Mabry, G.H. McKinley, R.E. Cohen, Robust omniphobic surfaces. Proc. Natl. Acad. Sci. **105**(47), 18200–18205 (2008)
14. X.-L. Meng, L.-S. Wan, Z.-K. Xu, Insights into the static and advancing water contact angles on surfaces anisotropised with aligned fibers: Experiments and modeling. Colloids Surf. A **389**(1–3), 213–221 (2011)
15. S. Koombhongse, W. Liu, D.H. Reneker, Flat polymer ribbons and other shapes by electrospinning. J. Polym. Sci. Part B: Polym. Phys. **39**(21), 2598–2606 (2001)
16. A. Cassie, S. Baxter, Wettability of porous surfaces. Trans. Faraday Soc. **40**, 546–551 (1944)
17. R.N. Wenzel, Resistance of solid surfaces to wetting by water. Ind. Eng. Chem. **28**(8), 988–994 (1936)
18. Y.-J. Sheng, S. Jiang, H.-K. Tsao, Effects of geometrical characteristics of surface roughness on droplet wetting. J. Chem. Phys. **127**(23), 234704 (2007)
19. G. Whyman, E. Bormashenko, How to make the Cassie wetting state stable? Langmuir **27**(13), 8171–8176 (2011)
20. L. Barbieri, E. Wagner, P. Hoffmann, Water wetting transition parameters of perfluorinated substrates with periodically distributed flat-top microscale obstacles. Langmuir **23**(4), 1723–1734 (2007)
21. A. Bussonnière, M.B. Bigdeli, D.-Y. Chueh, Q. Liu, P. Chen, P.A. Tsai, Universal wetting transition of an evaporating water droplet on hydrophobic micro-and nano-structures. Soft Matter **13**(5), 978–984 (2017)
22. J. Ou, G. Fang, W. Li, A. Amirfazli, Wetting transition on textured surfaces: a thermodynamic approach. J. Phys. Chem. C **123**(39), 23976–23986 (2019)
23. W. Li, A. Amirfazli, A thermodynamic approach for determining the contact angle hysteresis for superhydrophobic surfaces. J. Colloid Interface Sci. **292**(1), 195–201 (2005)
24. L. Tie, Z. Guo, W. Liu, Anisotropic wetting properties on various shape of parallel grooved microstructure. J. Colloid Interface Sci. **453**, 142–150 (2015)
25. L. He, W. Liang, Z. Wang, B. Yang, Z. Duan, Y. Chen, 3-D thermodynamic analysis on wetting behavior of superhydrophobic surfaces. Colloids Surf. A **504**, 201–209 (2016)
26. D. Kim, N.M. Pugno, S. Ryu, Wetting theory for small droplets on textured solid surfaces. Sci. Rep. **6**(1), 1–8 (2016)
27. J. Eick, R. Good, A. Neumann, Thermodynamics of contact angles. II. Rough solid surfaces. J. Colloid Interface Sci. **53**(2), 235–248 (1975)
28. R. Shuttleworth, G. Bailey, The spreading of a liquid over a rough solid. Discuss. Faraday Soc. **3**, 16–22 (1948)
29. J. Long, M. Hyder, R. Huang, P. Chen, Thermodynamic modeling of contact angles on rough, heterogeneous surfaces. Adv. Coll. Interface Sci. **118**(1–3), 173–190 (2005)
30. M. Reyssat, D. Quéré, Contact angle hysteresis generated by strong dilute defects. J. Phys. Chem. B **113**(12), 3906–3909 (2009)
31. R. Raj, R. Enright, Y. Zhu, S. Adera, E.N. Wang, Unified model for contact angle hysteresis on heterogeneous and superhydrophobic surfaces. Langmuir **28**(45), 15777–15788 (2012)
32. H.-J. Butt, C. Semprebon, P. Papadopoulos, D. Vollmer, M. Brinkmann, M. Ciccotti, Design principles for superamphiphobic surfaces. Soft Matter **9**(2), 418–428 (2013)
33. J. Panter, Y. Gizaw, H. Kusumaatmaja, Multifaceted design optimization for superomniphobic surfaces. Sci. Adv. **5**(6), eaav7328 (2019)
34. Z. Zhang, M. Zhao, Y. Ahn, J. Jang, Wettability of a surface engraved with the periodic nanoscale trenches: effects of geometry and pressure. J. Mol. Liq. **335**, 116276 (2021)

35. Y. Kaufman, S.-Y. Chen, H. Mishra, A.M. Schrader, D.W. Lee, S. Das et al., Simple-to-apply wetting model to predict thermodynamically stable and metastable contact angles on textured/rough/patterned surfaces. J. Phys. Chem. C **121**(10), 5642–5656 (2017)
36. A. Marmur, From hygrophilic to superhygrophobic: theoretical conditions for making high-contact-angle surfaces from low-contact-angle materials. Langmuir **24**(14), 7573–7579 (2008)
37. T. Young, III. An essay on the cohesion of fluids. Philos. Trans. R. Soc. Lond. **95**, 65–87 (1805)
38. P.-G. Gennes, F. Brochard-Wyart, D. Quéré, *Capillarity and Wetting Phenomena: drops, Bubbles, Pearls, Waves* (Springer, 2004)
39. E. Bormashenko, Wetting of flat and rough curved surfaces. J. Phys. Chem. C **113**(40), 17275–17277 (2009)
40. R.J. Vrancken, H. Kusumaatmaja, K. Hermans, A.M. Prenen, O. Pierre-Louis, C.W. Bastiaansen et al., Fully reversible transition from Wenzel to Cassie–Baxter states on corrugated superhydrophobic surfaces. Langmuir **26**(5), 3335–3341 (2010)
41. D. Arisawa, Y. Umetsu, A. Yoshizawa, C. Hill, J. Eastoe, F. Guittard et al., Controlling water adhesion on superhydrophobic surfaces with bi-functional polymers. Colloids Surf. A **616**, 126307 (2021)
42. C. Yuan, M. Huang, X. Yu, Y. Ma, X. Luo, A simple approach to fabricate the rose petal-like hierarchical surfaces for droplet transportation. Appl. Surf. Sci. **385**, 562–568 (2016)
43. J.W. Gose, K. Golovin, M. Boban, J.M. Mabry, A. Tuteja, M. Perlin et al., Characterization of superhydrophobic surfaces for drag reduction in turbulent flow. J. Fluid Mech. **845**, 560–580 (2018)

4 Developing "Rose-Petal" Superhydrophobicity Using Aligned Electrospun Fibers

Superhydrophobic surfaces with rose-petal effect, also referred to as petal-effect surfaces, show contact angles greater than 150° and high droplet adhesion, which can pin the droplets even by turning upside down [1]. The petal-effect surfaces have such potential applications as in droplet transportation [2], lab-on-chip device [3], single molecule spectroscopy [4].

Petal-effect surfaces can be fabricated using chemical etching [5], electroless galvanic deposition [6], plasma reactive ion etching [4], and electrospinning [7]. Among them, electrospinning is a versatile technology to develop sufficient surface roughness for petal-effect surfaces by fabricating micro- and nanofibers. In contrast to subtractive manufacturing, electrospinning has little destruction on the surface to be modified.

Detailed electrospinning-based methods for the fabrication of petal-effect surfaces have been introduced in Sect. 2.3. In brief, petal-effect surfaces are produced by either one-step electrospinning [7–10] or electrospinning followed by post treatments [11, 12]. The high contact angle, i.e., greater than 150°, is realized by the surface roughness generated by fibers or beads; the high adhesion is achieved by droplets penetrating into inter-fiber spaces, resulting in increased liquid-solid contact interface. However, a fundamental understanding of achieving fiber-based petal effect by controlling fiber morphology is still lacking.

The fiber morphology parameters, including fiber diameter, inter-fiber distance, and fiber orientation, can influence the surface wettability as reported by earlier experimental studies. For example, decreasing poly(caprolactone) (PCL) fiber diameter from 2.2 to 0.58 μm increases contact angle from 119° to 129° [13]; decreasing the inter-fiber distance of nylon-11 nanofibers by increasing the surface coverage rate of fiber from 0.7 to 20% increases contact angle from 70° to 124° [14]; compared to randomly distributed fibers,

Y. Zhang, *Superhydrophobicity of Surfaces Dressed by Electrospun Fibers*, Synthesis Lectures on Green Energy and Technology, https://doi.org/10.1007/978-3-031-55552-7_4

directionally aligned polystyrene (PS) fibers show greater contact angle hysteresis of 57° than 34° [15].

A few theoretical works have studied the wetting on fibrous surfaces while most of them are focused on the stability Cassie-Baxter (CB) state to guide the design of lotus effect. For instance, Rawal [16] built an analytical model relating fiber parameters with the stability of CB state for randomly distributed fibers; Bucher et al. [17] modeled the critical hydro-static pressure of the CB state for aligned fibers using numerical simulation. However, as reported by Meng et al. [18], droplets on fibers-dressed surfaces can be pinned by the free energy barriers in the Wenzel state, resulting in a higher contact angle than the ECA. Thus, the CB state may not be necessary in developing petal-effect surfaces.

Therefore, the objective of this research is to strategically approach petal effect based on the understanding of the influence of fiber morphology on the surface wettability. A strategic approach enables to achieve petal effect with less fibers than that by random deposition, so less impact, such as reduction of transparency and air permeability, is caused to the original surfaces. Aligned electrospun fibers are used herein for the simplicity in leveling the contact interface and quantifying fiber parameters. Accordingly, the rest of this chapter is organized as follows. Section 4.1 introduces the experimental methods for fabricating aligned fibers, and characterizing fiber morphology and surface wettability; a theoretical wetting model is built based on the integrated model in Chap. 3 to help understand the wetting behaviors on surfaces dressed by aligned fibers. Then Sect. 4.2 experimentally studies the influence of fiber parameters on the wettability of surfaces dressed by aligned fibers with explanation of mechanism based on the wetting model; petal-effect surfaces are developed considering the influence of fiber morphology.

4.1 Methods

4.1.1 Materials

Polystyrene (PS, MW = 280 kDa, ICA = 95° [18]) and N,N-Dimethylformamide (DMF, ACS reagent, ≥99.8%) are purchased from Sigma Aldrich, US. All chemicals are used as received without any purification. It is worth noting that this book selected PS as an example. Other polymers such as PVDF and PMMA may also be used to develop petal effect. Transparent tape, with a water contact angle measured to be 97°, is supplied from Wisdom Electronics Inc. Deionized water is used for contact angle measurement.

4.1.2 Fabrication of Aligned Fibers

Aligned fibers are fabricated as follows. First, 3.2 g of PS pellets is dissolved in 10 mL of DMF to prepare a 32% (w/v) PS/DMF solution. Then the polymer solution is transferred

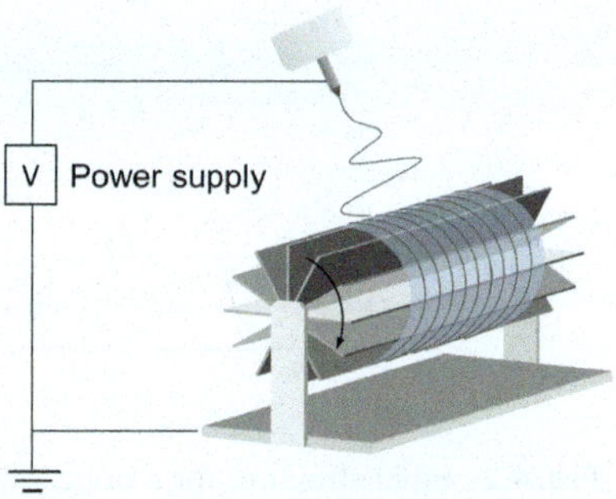

Fig. 4.1 Schematic of electrospinning with a rotating fin collector

to a syringe connected to a metallic needle (22 gauge) for electrospinning: the solution feeding rate is 0.8 mL/h; the applied voltage is 13.5 kV; the distance between collector and needle tip is 10 cm. A rotating fin collector, as shown in Fig. 4.1, with a rotating speed of 3100 rpm is used to align fibers. The alignment of fibers was driven by the mechanical stretching exerted by the rotating collector and the electrostatic force between the adjacent fins [19]. Transparent tapes are attached to the collector, so aligned fibers are deposited on the tapes in electrospinning. To fabricate fibers with a greater diameter, 36% (w/v) PS/DMF solution and accordingly voltage of 13.0 kV are used. The electrospinning time varies from 1 to 30 s to control inter-fiber distances. After electrospinning, fibers are dried in a vacuum oven to remove residual DMF.

4.1.3 Characterization of Aligned Fibers

The fiber morphology is characterized using a field-emission scanning electron microscope (Zeiss Leo 1530) at 5.0 kV. SEM images at magnifications of 1.0–2.0kX are analyzed by ImageJ, an image processing software, for fiber diameter. In addition, an optical microscope (Swift SW380B, USA) with a digital in-lens camera is used to capture images of the fibers at low magnifications of 40–1,000X. These images are then analyzed using Swift Imaging 3.0 software to determine the average center-to-center inter-fiber distance.

Then the apparent contact angles of the prepared fiber-dressed surfaces, with varying fiber diameters and inter-fiber distances, are measured using a contact angle goniometer (Ossila Ltd., UK) with sessile drop method: a 2 μL water droplet with standard deviation of 0.04 μL is generated by a 10 μL syringe (minimum graduation of 0.2 μL) with a gauge 32 stainless steel needle; then the 2 μL droplet is gently deposited onto the testing surfaces at a height of 1.5 mm; the measurement process is video recorded and the image for measuring contact angle is captured from the video 5 s after the droplet is deposited.

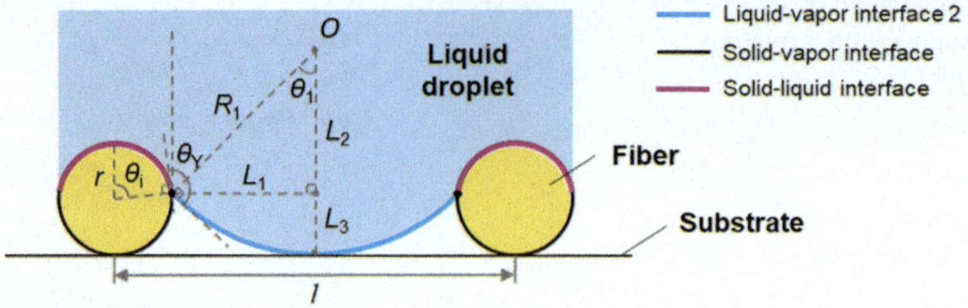

Fig. 4.2 Fiber structure for a noncomposite wetting state by sagging effect

4.1.4 Theoretical Analysis

A theoretical model is built based on Chap. 3 by treating fibers as circular structure with a smooth surface (or negligible surface roughness), to help understand the wetting behavior on surfaces dressed by aligned fibers. The wetting state of a droplet on a rough surface can be either composite or noncomposite. In this chapter, the droplet is expected to be in the noncomposite state because the inter-fiber distance, l, is much larger than the fiber radius, r. Specifically, Fig. 4.2 shows the geometry of fiber structure that causes a noncomposite state because of the sagging of underneath liquid-vapor interface onto the substrate. The relationship between r and l for a specific droplet volume is provided as follows.

To ensure a noncomposite state, the lowest point of the sagged liquid-vapor interface in Fig. 4.2 should have touched the bottom substrate even if there is no external perturbation to the three-phase system. Therefore, the following relationship should hold,

$$L_3 = r + r\cos\theta_i \tag{4.1}$$

where L_3 is the sagging depth of liquid-vapor interface, and θ_i is the angle representing the location of internal three-phase contact line (TPCL) on fiber, as shown in Fig. 4.2.

Based on the geometry relationship in Fig. 4.2, we can obtain Eqs. (4.2) and (4.3).

$$L_3 = R_1 - \sqrt{R_1^2 - L_1^2} \tag{4.2}$$

$$\theta_i = \pi - \theta_Y + \arcsin(L_1/R_1) \tag{4.3}$$

where R_1 is the radius of the sagged liquid-vapor interface; θ_Y is the intrinsic contact angle of fiber; L_1 is the horizontal distance between the internal TPCL and the lowest point of the sagged interface. Substituting Eqs. (4.2) and (4.3) into (4.1) gives,

$$R_1 - \sqrt{R_1^2 - L_1^2} = r\left(1 + \cos\left(\pi - \theta_Y + \arcsin(L_1/R_1)\right)\right) \tag{4.4}$$

Solving Eq. (4.4) gives the L_1, which can be substituted into Eqs. (4.3) and (4.5) for l.

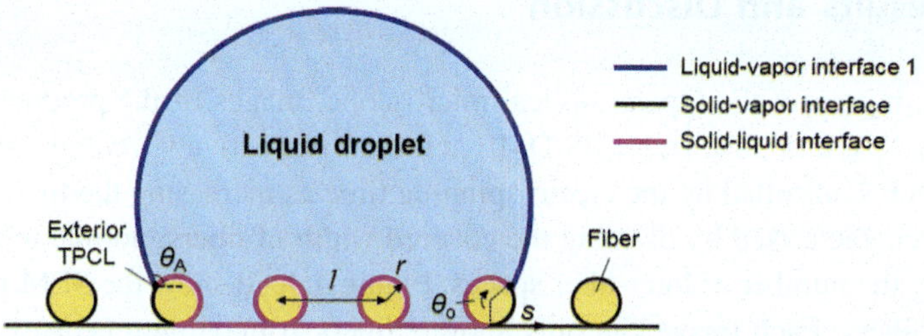

Fig. 4.3 Schematic of a droplet on a surface dressed by aligned fibers in noncomposite state

$$L_1 = l/2 - r\sin\theta_2 \tag{4.5}$$

Then the relationship between r and l for noncomposite state is obtained.

Figure 4.3 shows the schematic of a liquid droplet on a surface dressed by aligned fibers in noncomposite state. The fiber can be treated as the circular structure in Fig. 3.1(B-1).

For a surface patterned by fibers with an inter-fiber distance of l and a fiber radius of r, Eq. (3.4) can be specified into Eq. (4.6).

$$S = \frac{(\theta_A - \sin\theta_A\cos\theta_A)R_2^2}{\sin^2\theta_A} + (2n-1)r[l(1-\cos\theta_o) - r\pi] + r^2(\pi - \theta_o + \sin\theta_o\cos\theta_o) \tag{4.6}$$

where S is the constant [20, 21] cross-section area of droplet; θ_A is the apparent contact angle (APCA); R_2 is half of the basal contact width and can expressed using n, θ_o, and s, which are used to locate the external TPCL, as follows.

$$R_2 = (2n-1)l/2 - r\sin\theta_o + s \tag{4.7}$$

Then Eq. (3.8) can be specified into Eq. (4.8).

$$\frac{\Delta E}{\gamma_{LV}} = 2R_2\frac{\theta_A}{\sin\theta_A} - [4(n-1)\pi r + 2r\theta_o]\cos\theta_{Y1} - [(2n-1)l + 2s]\cos\theta_{Y2} \tag{4.8}$$

where E is the free energy of the three-phase system per unit depth; γ_{LV} is the liquid-vapor surface tension; θ_{Y1} and θ_{Y2} are the intrinsic contact angles of the fibers and the substrate, respectively. Equations (4.6–4.8) are the wetting model that relates free energy to the movement of the exterior TPCLs, which relates to APCA, on a surface dressed by aligned fibers.

4.2 Results and Discussion

Figure 4.4a, b present the typical optical microscope images of the produced aligned PS fibers using 32% and 36% (w/v) PS/DMF solution, respectively. Average inter-fiber distance, which is controlled by the electrospinning time as increasing the time decreases the distance, is characterized by dividing the covered width of fibers, normal to the alignment direction, by the number of inter-fiber spaces. Figure 4.4c, d show the SEM images for the aligned PS fibers. Each image can only show 2 fibers at maximum because of the sparse distribution. Then Fig. 4.4e, f provide the fiber diameter distribution by measuring more than 30 fibers for each concentration. As a result, 32% and 36% (w/v) solution generates beadless fibers with diameter of 0.9 ± 0.1 μm and 1.8 ± 0.2 μm, respectively, which are used to study the influence of fiber diameter on surface wettability. The solution viscosity increases with polymer concentration, resulting in thicker fibers; further increasing polymer concentration causes quick solidification of solution out of the spinning needle tip.

For fibers with diameter of 0.9 μm and 1.8 μm, the inter-fiber distance that ensures a 2 μL droplet, which is used in experiment, in noncomposite state is calculated to be greater than 39 μm and 56 μm, respectively, based on Eqs. (4.4) and (4.5). The inter-fiber distances in the following study are all greater than 70 μm, so droplets are in the noncomposite state.

Figure 4.5 shows the influence of fiber diameter (d) on the wettability of surfaces dressed by aligned fibers. We focus on that only two fibers are wetted by the droplet, such that the results are not interfered by the number and distribution of in-between fibers. The triangle and inverted triangle symbols denote the measured APCAs for fibers with d of 0.9 μm and 1.8 μm, respectively. The black dashed line shows the theoretical equilibrium contact angle (ECA), predicted using Eqs. (4.6–4.8) by locating the global minimum of free energy. The experimental APCAs deviate from the theoretical ECAs in Fig. 4.5, indicating that the droplets should be pinned at metastable states.

To explain the metastable state in Figs. 4.5 and 4.6 shows the variation of free energy per unit depth with APCA by substituting fiber parameters of d = 1.8 μm, l = 1060 μm (as an example to explain the metastable state), droplet volume of 2 μL, $\theta_{Y1} = 95°$, $\theta_{Y2} = 97°$ into Eqs. (4.6–4.8). The ECA corresponds to the APCA where the global minimum of free energy is located. As indicated by the inset of Fig. 4.6, an energy barrier, which can hinder the three-phase system from reaching the global minimum of free energy, is formed at the location where fibers are located. As a result, the droplet is pinned at the local minimum of free energy.

Based on Fig. 4.6, the predicted contact angles for the experimental fiber parameters are calculated using Eqs. (4.6–4.8) at the first energy barrier, since only two fibers wetted, and plotted as the solid lines in Fig. 4.5. The prediction fits with the experimental data with a mean difference of 2.9%. Then the influence of d is discussed as follows.

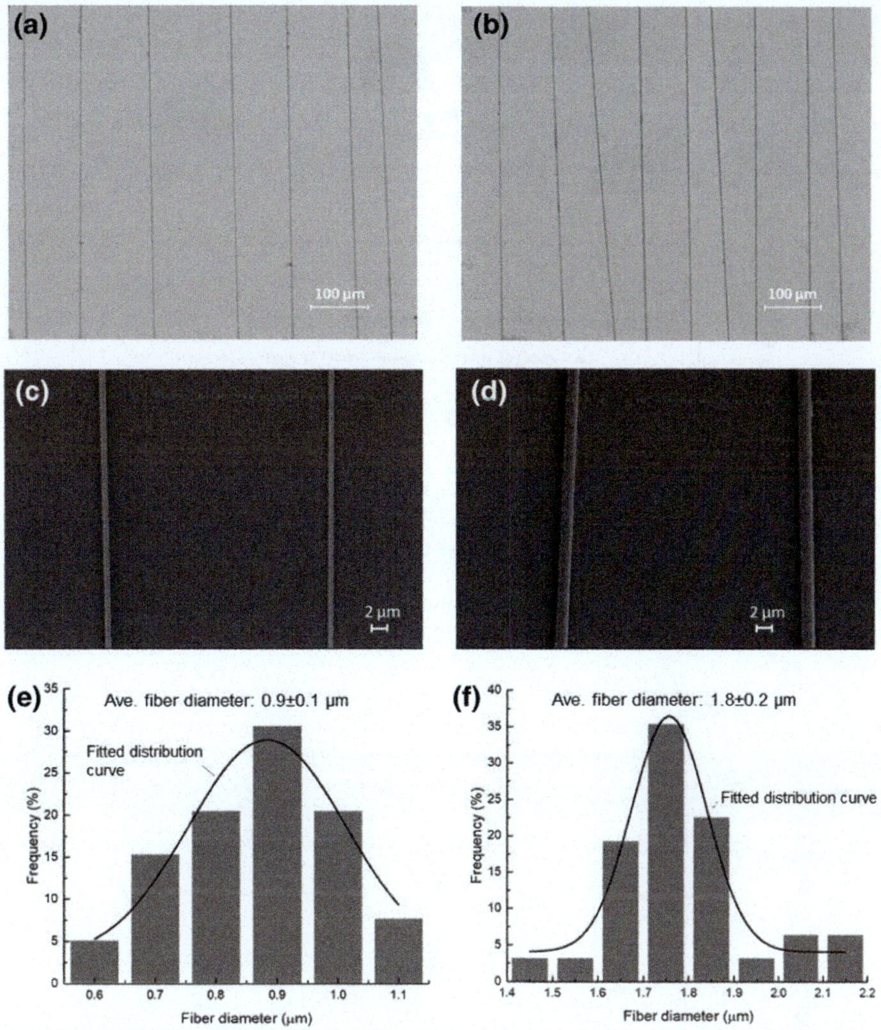

Fig. 4.4 **a**, **b** Optical microscope images (40X magnification) of the fabricated aligned PS fibers using (A) 32% and (B) 36% (w/v) PS/DMF solution; **c**, **d** SEM images (2.0 kX magnification) of (C) 32% and (D) 36% (w/v); **e**, **f** fiber diameter distribution of **e** 32% and **f** 36% (w/v) by measuring more than 30 fibers

In Fig. 4.5, when l is above 1060 μm, the APCAs for both fiber diameters increase with decreasing l if only the energy barrier can pin the droplet from further spreading. The same increasing pattern for both fiber diameters is attributed to that the d is much smaller than the scale of droplet (<2 μm vs. ~1600 μm), thus APCA is dominated by the basal width of droplet according to Eq. (4.6). However, the APCA of 0.9 μm fibers

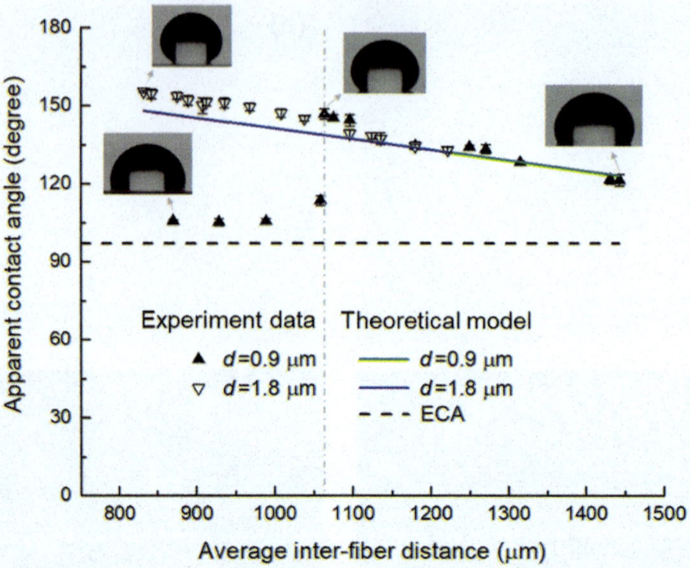

Fig. 4.5 Influence of fiber diameter on the wettability of surfaces dressed by aligned fibers (average fiber diameter: 0.9 and 1.8 μm; droplet volume: 2.0 μL)

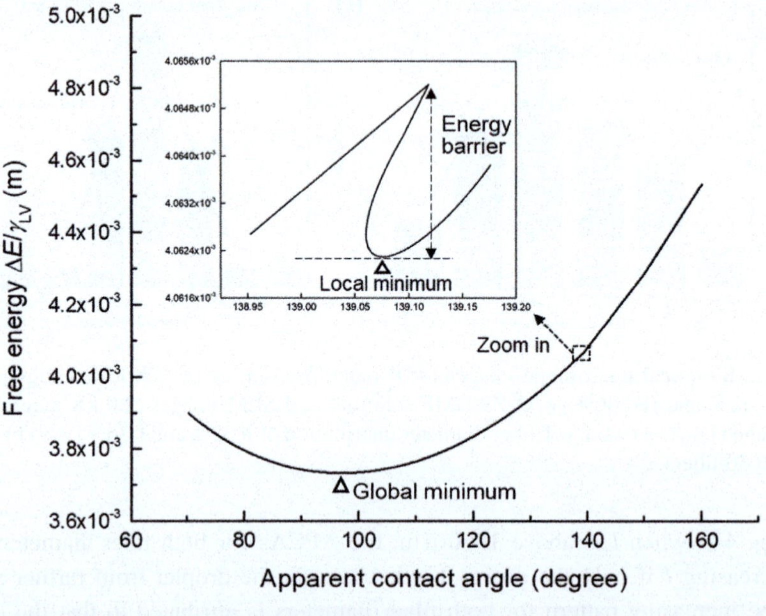

Fig. 4.6 Variation of free energy per unit depth with APCA and illustration of energy barrier (d = 1.8 μm; l_1 = 1060 μm; droplet volume: 2 μL; $\theta_{Y1} = 95°$; $\theta_{Y2} = 97°$)

falls to 113° and then 105° when l is below 1060 μm because the droplet overcomes the energy barrier and cross the two fibers. In contrast, the APCA of 1.8 μm fibers keeps increasing to 155° until $l = 830$ μm. The difference of the maximum APCAs between the two fibers indicates the difference of magnitude of energy barrier.

Figure 4.7 shows the variation of magnitude of free energy barrier with the basal width of droplet ($2R_2$) for different fiber diameters. The magnitude of energy barrier is calculated by subtracting the free energy at one metastable state from the following peak of free energy, as explained in the inset of Fig. 4.6. The basal width of droplet, equals to the l when only two fibers are wetted. As indicated in Fig. 4.7, the magnitude of energy barrier increases with the basal width and d. Specifically, the increment with basal width is essentially due to the increased ΔL_{LV1} as explained in Sect. 3.3. For the increment with d, the local minimum of free energy locates at the same angular position of fiber for different d based on Eq. (3.23), but fiber with larger diameter has a greater arc length along the spreading of droplets. Thus, the larger maximum APCA for 1.8 μm fiber in Fig. 4.5 results from the higher magnitude of energy barrier.

Figure 4.8 presents the influence of average inter-fiber distance (l) on the wettability of surfaces dressed by aligned fibers with an average fiber diameter of 1.8 μm. Specifically, as l decreases from 1000 μm to 210 μm, the number of fibers being wetted increases from 2 to 5. The triangle, inverted triangle, square, and round points denote the measured APCAs for wetting 2, 3, 4, and 5 fibers, respectively.

Pinning a droplet at a metastable state indicates that the perturbation energy to the droplet is not able to overcome the free energy barrier. At the point A of Fig. 4.8, where two fibers are wetted and the l is about 1000 μm, the energy barrier induced by the two fibers is greater than the perturbation energy, so the APCA of point A can be realized. As we decrease l, the magnitude of energy barrier decreases (see Fig. 4.7) while the perturbation energy remains nearly constant for a given measurement environment. At

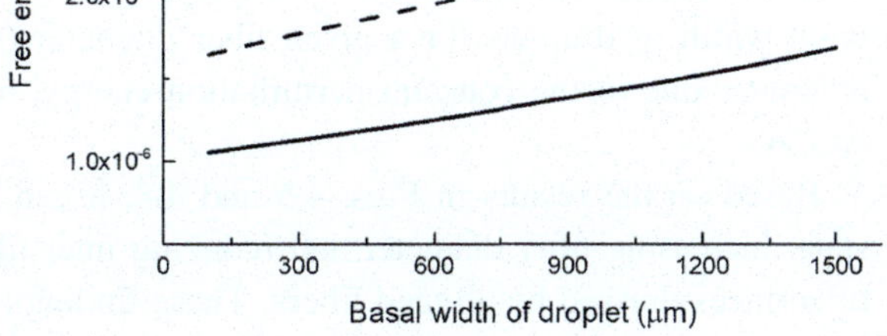

Fig. 4.7 Variation of magnitude of free energy barrier with the basal width of droplet for different fiber diameters (droplet volume: 2 μL; $\theta_{Y1} = 95°$; $\theta_{Y2} = 97°$)

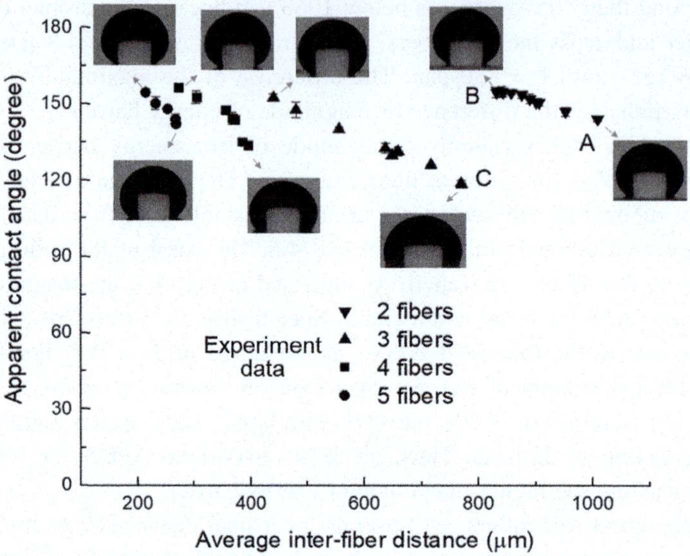

Fig. 4.8 Influence of inter-fiber distance on surface wettability (average fiber diameter: 1.8 μm; droplet volume: 2.0 μL)

point B ($l \sim 800$ μm), the energy barrier is very close to (or slightly greater than) the perturbation energy. Further decreasing l results in that the droplet overcomes the energy barrier and wets more fiber, i.e., jumping from point B to C. From B to C, the basal width of droplet increases from l to $2\,l$, resulting a higher energy barrier, but the APCA at point C drops below 120°.

In Fig. 4.8, similar variation pattern of APCA as the change from A–B to C can be observed for wetting three to five fibers. However, the overall APCAs for three to five fibers increase when l is below 800 μm. The reason is attributed to the decreased l from three to five fibers: once the energy barrier pinning the droplet is overcome, the following fiber that can pin the droplet becomes closer. In other words, more energy barriers are induced by the denser fibers to attenuate the perturbation energy to the droplet. In addition, the maximum APCAs for wetting three to five fibers are nearly fixed around 155° because of the nearly constant perturbation energy to droplets. At the maximum APCA, the perturbation energy is very close the energy barrier, which is only determined by the basal width of droplets for a given fiber diameter (see Fig. 4.7). Thus, the basal width is similar due to the constant perturbation energy, resulting in a nearly fixed maximum APCA.

Based on the results in Figs. 4.5 and 4.8, it can be concluded that, in noncomposite state, increasing fiber diameter and reducing inter-fiber distance can increase the APCA of surfaces dressed by aligned fibers. These findings can guide the development of petal-effect surfaces in the next. However, the fiber diameter here is hard to be further increased because ensuring fiber alignment needs a high rotating speed of the electrospinning collector, which reduces fiber diameter because of the extra stretching of fibers on collector. Therefore, we reduce the inter-fiber distance for petal effect.

Figure 4.9a shows the APCA of fibers (d = 1.8 μm) with inter-fiber distance l varying from 70 to 1000 μm. When l lies between 160 and 1000 μm (solid inverted triangles), less than eight fibers are wetted and the APCA varies in the range of 118°–156°. As l decreases below 130 μm (solid triangles), eight or more fibers are wetted, resulting in an APCA range of 150°–156°. The stable APCA for $l < 130$ μm is attributed to the multiple free energy barriers induced before the pinning location. Moreover, the same highest APCA of 156° for l varying in 160–1000 and 70–130 μm is determined by the environment perturbation.

Furthermore, the adhesion of fiber-dressed surfaces (d = 1.8 μm and $l < 130$ μm) to water droplet is characterized by tilting the surfaces after deposition of water droplets. As a result, droplets can adhere to all the surfaces by turning the surfaces upside down. Figure 4.9b shows that the high adhesion enables to turn the fiber-dressed surface with a droplet having a contact angle of 156° upside down. After turning back, the basal width and the APCA of the droplet remain unchanged. The high adhesion is attributed to liquid droplet penetrating into the spaces between adjacent fibers, resulting in a high liquid–solid contact area.

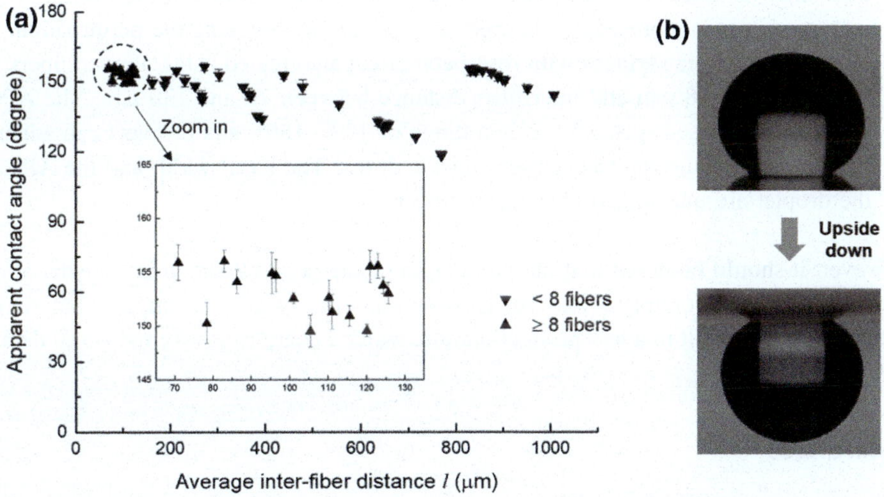

Fig. 4.9 **a** The apparent contact angle of fiber-dressed surfaces with inter-fiber distance varying from 70 to 1000 μm (average fiber diameter: 1.8 μm; droplet volume: 2.0 μL); **b** Illustration of rose petal effect by turning the surface with a droplet upside down

4.3 Summary

This research studied the influence of fiber parameters on the wettability of surfaces dressed by aligned fibers and developed petal-effect surfaces with a strategic approach. Based on this research, the following conclusions can be reached.

(1) A thermodynamic model relating free energy with the apparent contact angle (APCA) on the surfaces dressed by aligned fibers is built. Free energy barriers can pin a droplet at metastable states, resulting in higher APCAs than the equilibrium contact angle. The model indicates that the magnitude of energy barrier increases with the fiber diameter and basal width of droplets.

(2) Aligned polystyrene fibers with diameter of 0.9 and 1.8 μm and varying inter-fiber distance from 80 to 1450 μm are produced using one-step electrospinning. At large inter-fiber distance, >800 μm, where two fibers can pin a 2 μL droplet from spreading to equilibrium state, the maximum APCA of the 1.8 μm fibers is 155°, compared to that of 147° for 0.9 μm fibers. The greater APCA is due to the higher magnitude of energy barrier for 1.8 μm fibers.

(3) When the inter-fiber distance is below 800 μm, two 1.8 μm fibers are unable to pin the droplet because of the reduced energy barrier with inter-fiber distance, so three fibers are wetted with APCA reducing below 120°. Similar variation of APCA by reducing inter-fiber distance are observed for three to five fibers. However, the overall APCAs increase because more energy barriers are induced to attenuate the perturbation.

(4) Superhydrophobic surfaces with rose-petal effect are created using aligned fibers with a diameter of 1.8 μm and inter-fiber distance between 70 and 130 μm. The APCAs of the fiber-dressed surfaces are in a range of 150°–156°, and droplets can adhere to the surfaces by turning the surfaces upside down. The basal width and the APCA of the droplet are unchanged after turning back.

However, it should be noted that the fabrication of the petal effect surfaces using aligned fibers is still in a laboratory scale. The rotating fin collector and the single spinneret used in this book can result in a low production rate, which is hard to satisfy industrial demand.

References

1. D. Ebert, B. Bhushan, Wear-resistant rose petal-effect surfaces with superhydrophobicity and high droplet adhesion using hydrophobic and hydrophilic nanoparticles. J. Colloid Interface Sci. **384**(1), 182–188 (2012)
2. X. Hong, X. Gao, L. Jiang, Application of superhydrophobic surface with high adhesive force in no lost transport of superparamagnetic microdroplet. J. Am. Chem. Soc. **129**(6), 1478–1479 (2007)

3. A.I. Neto, C.R. Correia, C.A. Custódio, J.F. Mano, Biomimetic miniaturized platform able to sustain arrays of liquid droplets for high-throughput combinatorial tests. Adv. Func. Mater. **24**(32), 5096–5103 (2014)
4. A. Winkleman, G. Gotesman, A. Yoffe, R. Naaman, Immobilizing a drop of water: fabricating highly hydrophobic surfaces that pin water droplets. Nano Lett. **8**(4), 1241–1245 (2008)
5. H.T. Bahrami, B. Ahmadi, H. Saffari, Preparing superhydrophobic copper surfaces with rose petal or lotus leaf property using a simple etching approach. Mater. Res. Express **4**(5), 055014 (2017)
6. C. Yuan, M. Huang, X. Yu, Y. Ma, X. Luo, A simple approach to fabricate the rose petal-like hierarchical surfaces for droplet transportation. Appl. Surf. Sci. **385**, 562–568 (2016)
7. G. Gong, J. Wu, J. Liu, N. Sun, Y. Zhao, L. Jiang, Bio-inspired adhesive superhydrophobic polyimide mat with high thermal stability. J. Mater. Chem. **22**(17), 8257–8262 (2012)
8. F. Mikaeili, P.I. Gouma, Super water-repellent cellulose acetate mats. Sci. Rep. **8**(1), 12472 (2018)
9. H. Yoshida, K. Yanagisawa, Creation of superhydrophobic poly (L-phenylalanine) nonwovens by electrospinning. Polymers **10**(11), 1212 (2018)
10. H. Yoshida, D. Klee, M. Möller, M. Akashi, Creation of superhydrophobic electrospun nonwovens fabricated from naturally occurring poly (amino acid) derivatives. Adv. Func. Mater. **24**(40), 6359–6364 (2014)
11. G. Gong, J. Wu, X. Jin, L. Jiang, Adhesion tuning at superhydrophobic states: from petal effect to lotus effect. Macromol. Mater. Eng. **300**(11), 1057–1062 (2015)
12. T. Pisuchpen, N. Chaim-Ngoen, N. Intasanta, P. Supaphol, V.P. Hoven, Tuning hydrophobicity and water adhesion by electrospinning and silanization. Langmuir **27**(7), 3654–3661 (2011)
13. M. Ma, Y. Mao, M. Gupta, K.K. Gleason, G.C. Rutledge, Superhydrophobic fabrics produced by electrospinning and chemical vapor deposition. Macromolecules **38**(23), 9742–9748 (2005)
14. D. Bagrov, S. Perunova, E. Pavlova, D. Klinov, Wetting of electrospun nylon-11 fibers and mats. RSC Adv. **11**(19), 11373–11379 (2021)
15. A. Rawal, S. Shukla, S. Sharma, D. Singh, Y.-M. Lin, J. Hao et al., Metastable wetting model of electrospun mats with wrinkled fibers. Appl. Surf. Sci. **551**, 149147 (2021)
16. A. Rawal, Design parameters for a robust superhydrophobic electrospun nonwoven mat. Langmuir **28**(6), 3285–3289 (2012)
17. T. Bucher, M. Amrei, H.V. Tafreshi, Wetting resistance of heterogeneous superhydrophobic coatings with orthogonally layered fibers. Surf. Coat. Technol. **277**, 117–127 (2015)
18. X.-L. Meng, L.-S. Wan, Z.-K. Xu, Insights into the static and advancing water contact angles on surfaces anisotropised with aligned fibers: experiments and modeling. Colloids Surf. A **389**(1–3), 213–221 (2011)
19. P. Katta, M. Alessandro, R. Ramsier, G. Chase, Continuous electrospinning of aligned polymer nanofibers onto a wire drum collector. Nano Lett. **4**(11), 2215–2218 (2004)
20. L. Tie, Z. Guo, W. Liu, Anisotropic wetting properties on various shape of parallel grooved microstructure. J. Colloid Interface Sci. **453**, 142–150 (2015)
21. E. Bormashenko, Wetting of flat and rough curved surfaces. J. Phys. Chem. C **113**(40), 17275–17277 (2009)

5 Developing "Lotus" Superhydrophobicity Using Aligned Porous Fibers

Superhydrophobic surfaces with lotus effect, also termed as lotus-effect surfaces [1, 2], are featured with a high water contact angle (>150°) and a low roll-off angle (<10°) [3]. The lotus-effect surfaces have a variety of applications in developing functional materials including self-cleaning [4], anti-corrosion [5], anti-icing [6].

Surface roughness is necessary for developing a lotus-effect surface since a material with intrinsic contact angle greater than 120° is rarely available [7]. For the wetting on a rough surface, droplets can be in Wenzel [8] or Cassie-Baxter (CB) [9] state. Compared to the Wenzel state, droplets in the CB state have less contact area with solids, resulting in lower adhesion to the surfaces and thereby a lower roll-off angle. Therefore, CB state is preferable for creating lotus-effect surfaces.

Lotus-effect surfaces can be fabricated by a few techniques including lithography [10], chemical etching [11], laser micromachining [12], and electrospinning [13]. Among them, electrospinning is a facile and versatile technique to generate continuous micro- and nanofibers by applying a high voltage on a viscous polymeric solution. The generated thin fibers can create surface roughness, enabling electrospinning to be an efficient way for developing lotus-effect surfaces. In addition, different from methods such as chemical etching, laser micromachining that need to subtract materials from surface, electrospinning modifies surface wettability by introducing new materials, resulting in less impact on the original surface.

A few electrospinning-based methods are developed for producing lotus-effect surfaces, which have been introduced in Sects. 2.1 and 2.2 of this book. Briefly, these methods are classified into one-step electrospinning and electrospinning followed by post treatments. However, current electrospinning-based methods for developing superhydrophobicity often require complex fabrication processes and multiple raw materials,

Y. Zhang, *Superhydrophobicity of Surfaces Dressed by Electrospun Fibers*, Synthesis Lectures on Green Energy and Technology, https://doi.org/10.1007/978-3-031-55552-7_5

which increase the overall cost and hinder them from broad implementation. For example, synthesizing and applying low surface energy polymers into the electrospinning solution [14, 15], adding nanosized additives to the fiber surface [16, 17], and various post-treatments to incorporate a secondary material, such as PPFEMA [18, 19], DMDCS [20], and DTMS [21], with desired surface energy are commonly reported in the literature.

Moreover, there is a lack of a strategic approach to superhydrophobicity by electrospun fibers. Current theoretical works have limited ability in guiding the development of superhydrophobic electrospun structures because either the works are limited to only studying the stability of wetting state or the theoretical structures are difficult in replicating because of oversimplification of the fibrous structures in developing model. For example, Rawal [22] developed an analytical model relating fiber parameters with the pressure difference across the liquid-vapor interface. Based on the model, the influence of fiber diameter and inter-fiber distance on the CB stability is analyzed. Moreover, Emami et al. [23] treated fibrous surface as a plane by neglecting the height difference of fibers in different layers to explore the influence of fiber orientation on the CB stability. In addition, Onda [24] built a model for the equilibrium contact angle (ECA) of fibrous surfaces. Fibrous structure is treated as a porous film, so the ECA is related to volume porosity, film thickness, etc.

Therefore, the objective of this study is to develop a superhydrophobic electrospun fibrous structure by a strategic approach based on the understanding of droplets wetting behavior on fibrous structures. Uniaxially and biaxially aligned porous polystyrene (PS) fibers are studied because such structure may assist in achieving the lotus effect, due to (1) the porous fiber surface reduces the contact area between fibers and droplets (since the nanopores on the fiber surface can resist liquid from penetration due to the intrinsic hydrophobicity of PS and the closed pore [25]), facilitating the movement of droplet by reducing the resistance; (2) the alignment serves as a track to facilitate droplet rolling at a low tilting angle.

Accordingly. the rest of this chapter is organized as follows. Section 5.1 first introduces the experimental methods for fabrication of uniaxially and biaxially aligned porous fibers, and characterizations of fiber morphology and hydrophobicity. Then Sect. 5.1.4 builds a numerical model using Surface Evolver to determine the wetting state of droplets on the fiber-dressed surfaces. Section 5.2.1 provides the morphology of the fabricated fibers. The experimental wetting behaviors on the uniaxially and biaxially aligned porous fibers, shown in Sect. 5.2.2, are explained in Sect. 5.2.3 based on the numerical simulation. Then Sect. 5.2.4 develops a lotus-effect surface by a rational approach considering the effects of fiber-surface pores and liquid-fiber contact area. Finally, Sect. 5.3 summarizes this chapter.

5.1 Methods

5.1.1 Fabrication of Aligned Porous Fibers

Aligned porous fibers are produced by electrospinning as follows. First, a polymer solution is prepared by dissolving 1.9 g of polystyrene (PS, obtained from Sigma-Aldrich, US, with a molecular weight of 280 kDa and an intrinsic water contact angle of 95°) into 10 mL of tetrahydrofuran (THF, obtained from Sigma-Aldrich, US, ACS reagent grade $\geq$99%). The THF has a low boiling point at 66 °C, such that its quick evaporation in electrospinning can help to create pores on fiber surfaces by breath figure [26]. Then, the PS solution is loaded into a syringe with a 22 gauge metallic needle for electrospinning with the solution feeding rate of 4.4 ml/h, the applied voltage of 6 kV, and the needle tip to collector distance of 10 cm. A rotating fin collector covered with transparent tape is used to collect the uniaxially aligned fibers. The collector rotates at a speed of 450–1,000 r/min and the deposition time is between 30 and 270 s. The transparent tape has an intrinsic contact angle of 95°. The relative humidity (RH) of the air inside the electrospinning chamber is controlled between 40 and 80% by a portable humidifier (Sparoom, US) to create pores on fiber surfaces. Finally, the biaxial structures are produced by rotating the tape 90° for depositing the upper-layer fibers.

5.1.2 Characterization of Fiber Morphology

The morphology of the prepared electrospun fibers is characterized using a field-emission scanning electron microscope (SEM, Zeiss Leo 1530, Germany) at 5.0 kV. SEM images at low magnification of 500X and high magnifications of 10–20 kX are analyzed by ImageJ software to measure fiber diameter, and pore size and pore area fraction, respectively, of the produced fibers. In addition, an optical microscope with a digital in-lens camera (Swift SW380B, USA) is used to capture images of the produced fibers at low magnifications of 40–1,000X. These images are then analyzed using Swift Imaging 3.0 software to determine the average center-to-center inter-fiber distance.

5.1.3 Characterization of Surface Wettability

The contact angle (θ_A) and roll-off angles (α_r) of the fiber-dressed surfaces are measured using a contact angle goniometer (Model 250, ramé-hart instrument, US) with the sessile drop method. Deionized water droplets with a volume of 3 μL, unless otherwise specified, are generated by an automated dispensing system of the goniometer and mounted on the testing surfaces. The contact angles along the fiber direction are measured from the images captured by the ramé-hart DROPimage Advanced software. After that, the roll-off angle is determined by tilting the sample platform of the goniometer in the direction that droplets roll along the fiber direction. The contact angle and roll-off angle of each tested surface are measured 3–5 times, and average values are presented with standard deviation.

5.1.4 Development of Numerical Model

The model developed in this study determines the wetting state of water droplets on a fiber-dressed surfaces since the wetting behaviors of the droplet are closely related to the wetting state. A droplet can either fully or partially penetrate into the inter-fiber spaces, resulting in different contact angle and roll-off angle. Specifically, full penetration refers to the droplet touching the underneath substrate and filling all the inter-fiber spaces, i.e., in Wenzel state. Partial penetration means that the penetrated droplet stops at a certain layer of fibers and creates an air pocket between the bottom of the droplet and the underneath substrate. If the droplet stops at the top-layer fibers, it is noted as the Cassie-Baxter (CB) state.

The wetting state of a droplet on the fiber-dressed surface is controlled by its free energy *G*. The penetration of a liquid into the inter-fiber spaces may stop at a state where *G* is locally or globally minimized. The calculation of *G* is expressed as Eq. (5.1) [27].

$$\frac{G}{\gamma_{LV}} = A_{LV} - \iint_{A_{LF}} \cos\theta_{Y,F} dA - \iint_{A_{LS}} \cos\theta_{Y,S} dA \quad (5.1)$$

where A and γ are the surface area and surface tension, respectively; subscripts L, V, and F denote to liquid, vapor, and fiber, respectively; $\theta_{Y,F}$ and $\theta_{Y,S}$ are the intrinsic contact angles of the porous fiber and the substrate materials, respectively. The effect of gravity on droplet deformation is ignored since the radius of a 3 μL droplet (~0.89 mm) used in this study is smaller than the capillary length of ~2.7 mm [28].

Therefore, the free energy of the system, *G*, is essentially related to A_{LV}, A_{LF}, and A_{LS} as shown in Eq. (5.1). The latter three are variables changing with the droplet penetration and spreading on a fibrous structure. It brings challenges to analytically solve Eq. (5.1) because the liquid-vapor and liquid-fiber interfaces deform to complex geometries when the droplet is in contact with fibers (especially for multi-layer fibers). Thus, this study employs Surface Evolver (SE, version 2.70) finite element code to numerically solve Eq. (5.1) and predict the wetting state by minimizing the total free energy of the surface.

Figure 5.1 presents the schematic of the initial configuration of a water droplet on either uniaxially or biaxially aligned porous fibers for modeling in Surface Evolver. The droplet shape is initially set as cubic prior to minimizing total surface energy. The number of fibers in contact with the droplet is subject to the droplet volume. For a 3 μL droplet, the number of fibers contacted is between 4 and 30 based on the inter-fiber distance. It should be noted that only two fibers are shown in Fig. 5.1 to help visualize the contact

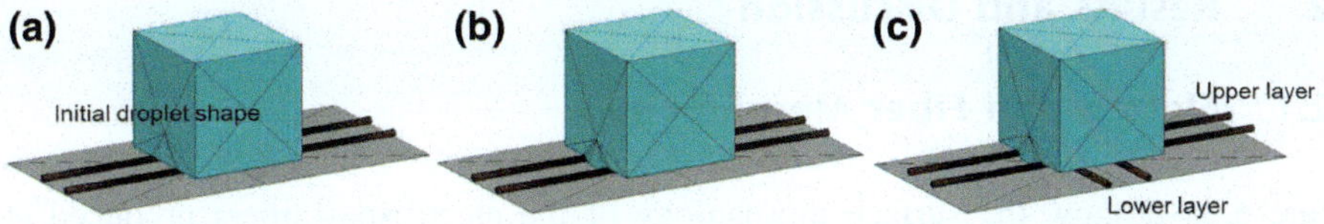

Fig. 5.1 Initial configuration of a water droplet contacting, **a** only fibers and **b** both fibers and substrate of uniaxially aligned porous fibers, and **c** the upper- and lower-layer fibers of biaxially aligned porous fibers for modeling using Surface Evolver

between droplet and fibers. Grid independence is checked by ensuring the variation of G is less than 0.01% after each successive refinement. For the biaxially aligned fibrous structure, the lower-layer fibers are set to have one fiber (or two for studying the inter-fiber distance of the lower layer). This configuration is deemed adequate for studying the trend of free energy changes during the droplet wetting, and it can also save computational costs by avoiding complicated mesh. In addition, the length and energy related parameters in the model are normalized to their dimensionless forms with respect to d_c, and $\gamma_{LV}\ d_c^2$ [29, 30], respectively. The d_c is set as 5 μm to represents the characteristic diameter of the produced fibers.

The effect of porous fibers on droplet wetting is approximately considered equivalent to that of using smooth fibers with an elevated θ_Y in the modeling since the SE is incapable of adding pores to fibers. In detail, the amendment of $\theta_{Y,F}$ is based on a strategy of formulating a non-porous surface that has the same energy change as the droplet does on a porous surface. For a droplet wetting a porous surface, the change of free energy ΔG is expressed as [27]

$$\frac{\Delta G}{\gamma_{LV}} = \Delta A_1 + (1 - f)\Delta A_2 - f\Delta A_2 \cos\theta_{Y,F} \tag{5.2}$$

where γ_{LV} is the interfacial tension between water and vapor phase; A_1 and A_2 are the surface areas of the outer surface of the droplet and the apparent contact area between the droplet and the substrate, respectively; f is the fraction of the actual surface wetted area over A_2; a porous surface with a higher porosity means a smaller f. In contrast, the energy change of a droplet wetting a non-porous surface is expressed as

$$\frac{\Delta G}{\gamma_{LV}} = \Delta A_1 - \Delta A_2 \cos\theta_{Y,F'} \tag{5.3}$$

where $\theta_{Y,F'}$ is the intrinsic contact angle of the non-porous surface. Then, Eqs. 5.2 and 5.3 give Eq. 5.4 when ΔG on the porous and non-porous surfaces are the same.

$$\theta_{Y,F} = \arccos\left(f \cos\theta_{Y,F'} - 1 + f\right) \tag{5.4}$$

5.2 Results and Discussion

5.2.1 Electrospun Fiber Morphology

Figure 5.2a–d show the surface morphology of the electrospun fibers produced at different RH. The results demonstrate the porous nature of fibers. The largest pore size of 180 ± 49 nm is obtained for electrospinning at the highest RH of 80%. As the RH reduces to 60%, 50%, and 40%, the pore size reduces to 160 ± 31 nm, 93 ± 22 nm, and 77 ± 16 nm, respectively. Furthermore, the fraction of the fiber surface occupied by pores (f_p) is related to the size of the pores. Thus, f_p also decreases from 51 to 34% with reducing RH for electrospinning. Additionally, it is also worth noting that increasing the speed of collector to more than 1000 rpm can alter the pore shape from approximate circle to oval because of the extra stretching to the fibers by the collector.

The formation of pores on fiber surfaces at a high RH fabrication environment of 40–80% is explained as follows [26, 31]. In electrospinning, the THF solvent in the ejected fibers evaporates to solidify the PS solute and form elongated fibers under an electric field. At the same time, the evaporation of THF is quick enough (due to its relatively low boiling point of 66 °C) to considerably reduce the fiber temperature, leading to the condensation of water vapor into droplets from the ambient air. Eventually, these condensed droplets result in the formation of the surface pores, because they occupy spaces before the ejected fiber is fully solidified (PS is hydrophobic and not soluble in water). This also explains the observation of bigger pore size and greater f_p for fibers fabricated at higher humidity, as it facilitates the condensation of water vapor onto the fiber surface during electrospinning. It is worth noting that the vapor-induced phase separation is not accounted for the pore formation because the compositional path of the PS solution stays in the homogeneous region in the ternary phase diagram of H_2O/THF/PS [31].

Figure 5.2e, f present the optical microscope images of the surfaces dressed by porous fibers, which shows that the fibers are directionally aligned. The average inter-fiber distance is obtained by dividing the covered width of fibers, normal to the alignment direction, by the number of inter-fiber spaces. In this study, the inter-fiber distance, ranging from 10 to 90 μm, is controlled by the electrospinning time as increasing the time decreases the distance.

5.2.2 Wetting Behavior of Surface Dressed by Aligned Porous Fibers

Figure 5.3 shows the preliminary experimental results for comparing the hydrophobicity of porous and non-porous fibers. The contact angle θ_A and roll-off angle α_r are measured on substrate surfaces (i.e., glossy side of an adhesive tape) dressed by uniaxially aligned fibers. The fiber diameters and inter-fiber distances of all samples are in

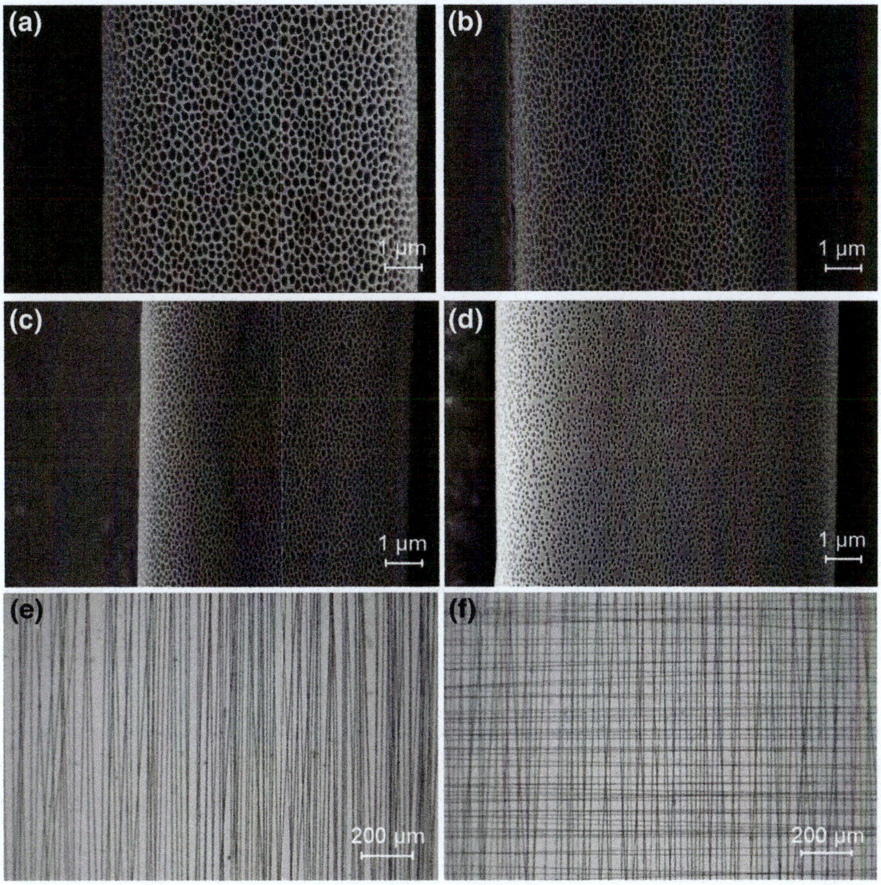

Fig. 5.2 SEM images of the aligned porous fibers electrospun at **a** 80%, **b** 60%, **c** 50%, and **d** 40% relative humidity, and optical microscope images of 40X magnification for surface dressed by **e** uniaxially and **f** biaxially aligned porous fiber

the range of 6.2–8.9 µm and 25.3–29.5 µm, respectively. Maintaining these two parameters within a relatively narrow range is advantageous in highlighting the influence of pores on hydrophobicity.

The results in Fig. 5.3 show the advantage of porous fibers in improving hydrophobicity. Increasing the average pore size from 0 to 180 ± 49.0 nm (corresponding to increasing the fraction of the fiber-surface area occupied by pores from 0 to 34–51%), leads to increasing contact angle from $134 \pm 2°$ to $152 \pm 1°$ and decreasing roll-off angle from $54 \pm 4°$ to $30 \pm 3°$. This improvement in hydrophobicity is attributed to the increase of the total liquid-vapor contact area when the droplet contacts fibers with greater pore area. Then, the observed θ_A increase can be explained by the CB equation [9]:

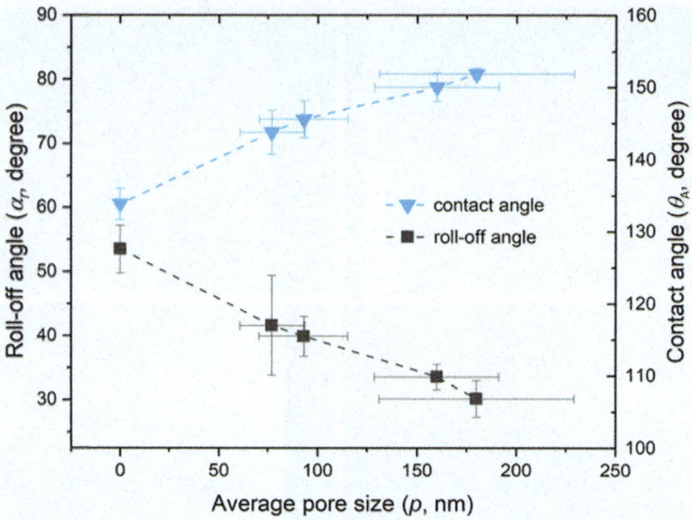

Fig. 5.3 Comparison of hydrophobicity of surfaces dressed by uniaxially aligned fibers with and without pores

$$\cos\theta_A = f_1 \cos\theta_Y - f_2 \tag{5.5}$$

where f_1 and f_2 represent the area fractions of the liquid-solid and liquid-vapor interfaces, respectively, over the unity parallel plane area of a rough surface. θ_Y is the intrinsic contact angle of the solid material. In this study, the sum of f_1 and f_2 is assumed to equal to 1, given that the cross section of the fiber is essentially an oval-shape with the fiber top having a relatively smaller curvature. This simplification is also adopted by other researchers [32, 33], who assume the fiber top as a plane for simplicity. Then, Eq. (5.5) is simplified as:

$$\cos\theta_A = -1 + f(1 + \cos\theta_Y) \tag{5.6}$$

where f is the liquid–solid contact area faction and can be calculated by

$$f = Nd(1 - f_p)/L_t \tag{5.7}$$

where N is the number of fibers across the measured area; d is the average fiber diameter, f_p is the pore area fraction on fiber surfaces; L_t is the total width from the first to the Nth measured fiber. The introduction of fiber pores reduces f and ultimately results in an increase in θ_A, as indicated by Eq. (5.6). The decrease in f also explains the observed decrease in α_r because liquid contacts more vapor phase, which reduces the friction that impedes droplet from rolling. The results in Fig. 5.3 concur with the earlier works [18,

32] of developing porous surfaces to improve hydrophobicity. So, all fibers are porous in the following studies.

Then the hydrophobicity of uniaxially and biaxially aligned porous fibers are compared. Figure 5.4 shows the measured contact and roll-off angles on surfaces dressed by uniaxially (see grey scatters) and biaxially (blue scatters) aligned porous fibers with all fibers having an average diameter of 3.95 ± 0.25 μm. The results show that increasing inter-fiber distance (l) from about 9–18.2 μm in both structures improves the hydrophobicity, as evidenced by the increased θ_A and decreased α_r from about 145° and 55° to 150° and 35°, respectively. The increased θ_A with enlarging l suggests a CB wetting state. Specifically, in the CB state, droplets are supported by fibers without touching the bottom substrate. As a result, increasing inter-fiber distance in CB state reduces the liquid-solid contact fraction f, thereby increasing the θ_A (see Eq. 5.6). Moreover, this decrease in f also explains the observed decrease in α_r because decreasing f reduces the liquid-solid contacting fraction, which reduces the friction that impedes droplet from rolling. In the rest of this study, this correlation between θ_A, α_r, and f is used to interpret changes in hydrophobicity for fiber-dressed surfaces before liquid wets the substrate (i.e., hydrophobicity improves with decreasing f, and vice versa).

The wetting behavior of droplets on these two fibrous structures diverges when the inter-fiber distance (l) is above 18.2 μm. For the uniaxial structure, further increasing l from 18.2 to 33.5 μm weakens the hydrophobicity, as evidenced by the decrease of θ_A from 155° to 130° and the increase of α_r from 32° to a sticky state even at 90° tilting

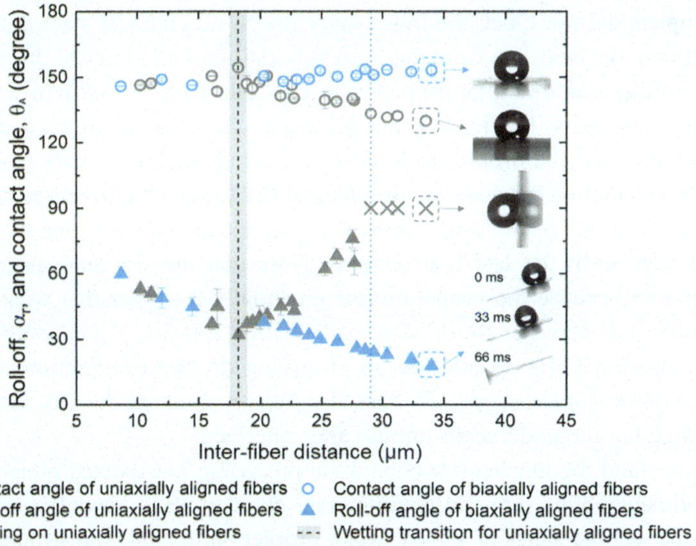

Fig. 5.4 Measured contact angle and roll-off angle on surfaces dressed by uniaxially or biaxially aligned porous fibers (water droplets with volume of 3 μL)

angle. This result indicates that a wetting transition occurs at $l > 18.2$ μm for the surfaces dressed by uniaxially aligned porous fibers: at this l, the liquid starts filling up some of many inter-fiber grooves under the droplet; further increasing l leads to more grooves filled by liquid and eventually results in the Wenzel state when the droplet fully wets the underlying substrate. The wetting transition weakens hydrophobicity since the liquid-solid contact fraction greatly increases once the droplet touches the substrate, enhancing the adhesion between the droplet and the covered area. In contrast, this wetting transition is not observed for droplets on the biaxially aligned porous fibers for l between 18.2 and 34 μm: the hydrophobicity continues to improve without a sudden change for all the l tested.

The experimental results are unexpected for the biaxial structure in Fig. 5.4. Note that the biaxial structure is essentially a stacking of two layers of uniaxially aligned fibers in the warp and weft directions, and therefore the droplets on the biaxial structure are expected to behave in the same way as on the uniaxial structure before reaching the lower-layer fibers. One would then intuitively expect that the wetting of the biaxial structure would transit to a state where the droplet contacts the lower-layer fibers once the l is greater than 18.2 μm. The reason is that at this l, the same droplet on the uniaxial structure would have touched the substrate. Had the droplet touched lower-layer fibers, the measured θ_A and α_r should have suddenly dropped and risen, respectively, because of the increase in liquid-solid contact fraction f. This weakening of hydrophobicity is expected to occur, even if it may not be as severe as seen in the uniaxial structure when $l > 18.2$ μm. However, the hydrophobicity of the biaxial structure maintains an unchanged trend for the tested l range of 8.7–34 μm. This unchanged relationship suggests that the liquid droplets did not touch the lower-layer fibers, i.e., the CB state is maintained.

In addition, the evidence from the free evaporation of an identical droplet on the biaxial structure further confirms that droplet is in CB state for $l > 18.2$ μm without contacting the lower-layer fibers. In Fig. 5.5, the deformations of liquid edges due to the capillary force evidence that the droplet is in contact with upper-layer fibers (see the yellow dot lines). The interaction between the droplet and the lower-layer fibers is not observed until the last few seconds of the evaporation (the red dot lines). Thus, the evaporating droplet is in CB state until the last few seconds. Note that the CB state is more stable prior to evaporation because the evaporation is essentially a trigger that weakens the stability of CB state and increases the tendency of wetting transition (i.e., evaporation-triggered wetting transition [34]). In addition to observing droplet evaporation, an inverted laser scanning confocal microscope [35] can also help to confirm the CB state of droplets by directly imaging the underneath liquid-vapor interface.

To understand the mechanism behind the preceding unexpected phenomenon, we map our experiment points into the wetting state chart of the uniaxial structure. Figure 5.6 presents the wetting states of a 3 μL water droplet on the uniaxial structure with different inter-fiber distances (l). The Cassie-Baxter region, i.e., blue region in Fig. 5.6, is obtained as follows.

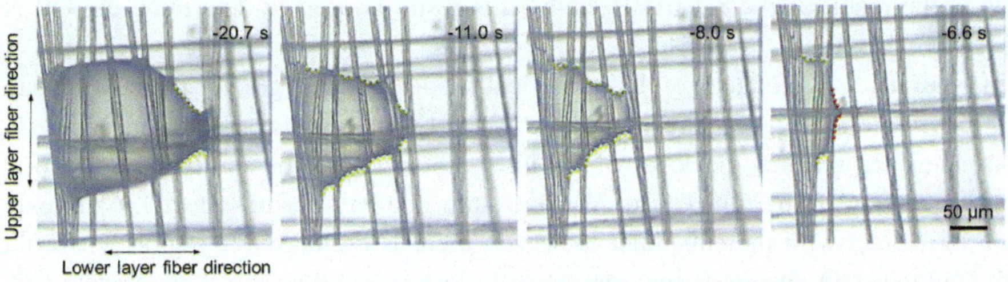

Fig. 5.5 Images of the final seconds of droplet free evaporation on the biaxially aligned porous fibers

Fig. 5.6 Experimental data points mapped in a wetting state chart obtained by modeling the uniaxially aligned fibrous structure with a fiber diameter of 3.95 μm

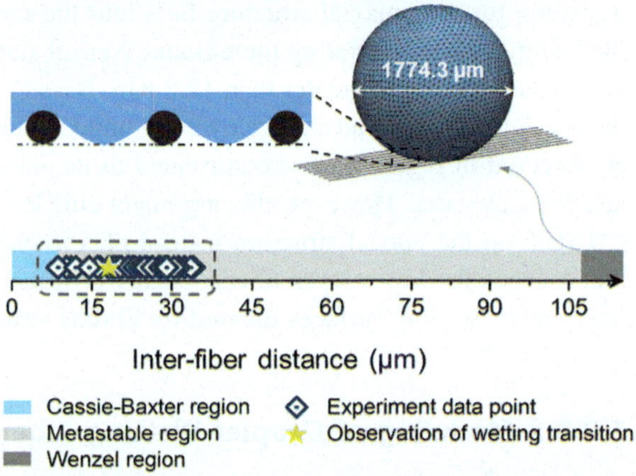

In wetting transition, the change of free energy (ΔG) can be expressed as follows based on Eq. (3.14).

$$\frac{\Delta G}{\gamma_{LV}} = -\left[l\left(1 + \cos\theta_{Y,S}\right) - d\sin\theta_{Y,F} - d(1 - f)\theta_{Y,F} + df\theta_{Y,F}\cos\theta_{Y,F}\right] \tag{5.8}$$

where $\theta_{Y,F}$ and $\theta_{Y,S}$ denote the intrinsic contact angles of the fiber and the substrate material, respectively; d is the fiber diameter, and f is the liquid-solid contact area fraction. The CB state dominates when it is thermodynamically more stable than the Wenzel state, i.e., $\Delta G > 0$, which derives Eq. (5.9).

$$l < \frac{d\left(\sin\theta_{Y,F} + (1 - f)\theta_{Y,F} - f\cos\theta_{Y,F}\right)}{1 + \cos\theta_{Y,S}} \tag{5.9}$$

Then, substituting the structure parameters and the material properties of the developed fibers in this study (i.e., $d = 3.95$ μm, $f = 0.5$, $\theta_{Y,F} = 95°$, $\theta_{Y,S} = 95°$) into Eq. (5.9) leads to the result of $l < 8.2$ μm. Finally, this result is used to determine the CB region in Fig. 5.6.

In contrast, as simulated by the numerical model in SE, l greater than 107.4 μm (the dark grey region in Fig. 5.6) results in the Wenzel state because the bottom of the liquid-vapor interface reaches the level of the fiber bottom (see the inset in Fig. 5.6). Thus, the wetting transition from CB to Wenzel state due to sagging effect [36] is expected when in contact with the substrate. However, there is a metastable region where the CB state could sustain despite being energetically less stable than the Wenzel state [37, 38]. The metastable CB state occurs because the additional energy acquired by the droplet from environment cannot overcome the free energy barrier between the CB and Wenzel states [39] (as discussed in Sect. 3.1.4). In Fig. 5.6, the experimentally observed wetting transition for the uniaxial structure falls into the metastable region, which indicates that the energy barrier separating the CB and Wenzel states is overcome by the environmental disturbances once l is greater than 18.2 μm. Based on this analysis, it can be inferred that the biaxial structure's advantage over the uniaxial structure in resisting wetting transition, as observed in Fig. 5.4, can be attributed to its enhanced energy barrier between the CB and Wenzel states. However, this argument still leaves a question of why a droplet is in CB state on the biaxial structure instead of a transitional state between CB and Wenzel that touches the lower-layer fibers. Then we further map a more detailed energy profile for droplets wetting surfaces dressed by fibrous structures in the following section.

5.2.3 Modeling of Droplet Wetting Fiber-Dressed Surface

The model parameters of droplet volume, fiber diameter, and pore area fraction on the fiber-surface are 3 μL, 4 μm, and 0.5, respectively, which are consistent with those in preceding experiment of Fig. 5.4; inter-fiber distance l is set as 30 μm because the wetting behaviors of droplets on the biaxial and uniaxial structures are notably different for $l >$ 29 μm (shown as the blue dotted line in Fig. 5.4, droplets are pinned on the uniaxial structure while roll off the biaxial structure). The amended $\theta_{Y,F}$ for the PS porous fibers is calculated to be 123° using Eq. (5.4).

Based on the model, we particularly studied the free energy of droplets in CB, Wenzel, and two transitional states to map the energy profile for droplet wetting fibrous structures. Figure 5.7 shows the variation of free energy in wetting surfaces dressed by aligned porous fibers. As illustrated in Fig. 5.7a, State(i) is the transitional state in which the droplet just starts to encounter the solids beneath the upper-layer fibers. The solid is the substrate surface for the uniaxial structure or the lower-layer fibers for biaxial structure. States(ii) represents any transitional states between State(i) and Wenzel state, e.g., State(ii′) and (ii″), in which the liquid partially wets the beneath solid. The normalized free energy of the droplet at CB, Wenzel and two transitional state is denoted as $\overline{G}_{\mathrm{CB}}$, $\overline{G}_{\mathrm{W}}$, $\overline{G}_{\mathrm{i}}$ and $\overline{G}_{\mathrm{ii}}$, respectively.

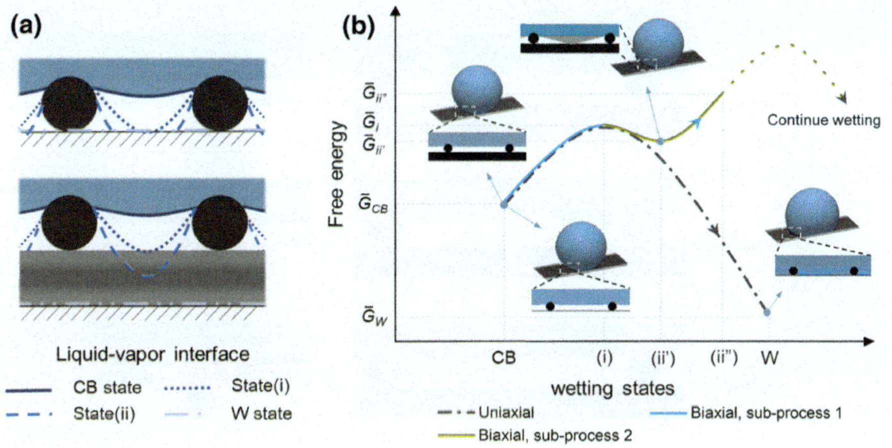

Fig. 5.7 **a** Illustration of four different wetting states. **b** Energy profile for a water droplet wetting uniaxial and biaxial structures, the insets are the wetting states with a local minimum of free energy obtained from modeling

The grey dash-dot line in Fig. 5.7b illustrates the mapped energy profile derived for wetting the surface dressed by uniaxially aligned porous fibers. The profiled is mapped as $\overline{G}_{\mathrm{i}} > \overline{G}_{\mathrm{CB}} > \overline{G}_{\mathrm{W}}$ because the model results identify CB and Wenzel as the wetting states with the lowest free energy before and after the droplet touching the substrate, respectively. The normalized free energy (with respect to $\gamma_{\mathrm{LV}}\ d_{\mathrm{c}}^{2}$) difference between CB and Wenzel states is 15,176.4. In addition, the free energy level of State(i) is at the peak because State(i) is a transitional state with the largest liquid-vapor interface (see Fig. 5.7a and Eq. 5.1). This energy profile concurs with the experimental results in Fig. 5.4 in which the Wenzel state occurs on the uniaxial structure with an inter-fiber distance of 30 μm because $\overline{G}_{\mathrm{W}}$ is the lowest among all wetting states.

However, as shown in Fig. 5.7b, the mapped energy profile for wetting the biaxial structure is different from wetting the uniaxial structure. The profile is mapped as $\overline{G}_{\mathrm{i}} > \overline{G}_{\mathrm{ii}} > \overline{G}_{\mathrm{CB}}$ by dividing the wetting into two consecutive sub-processes, where droplet wets only the upper-layer fibers (i.e., sub-process 1), and both the upper and lower-layer fibers (i.e., sub-process 2). The sub-process 1 is same as the process wetting the uniaxial structure before liquid touches substrate, thereby, we can map $\overline{G}_{\mathrm{i}} > \overline{G}_{\mathrm{CB}}$. For the sub-process 2, the modeling results show that there is a metastable state, i.e., State (ii′), soon after State(i) with a local minimum $\overline{G}_{\mathrm{ii'}}$, thereby, $\overline{G}_{\mathrm{i}} > \overline{G}_{\mathrm{ii'}}$. The obtained $\overline{G}_{\mathrm{ii'}}$ is calculated to be greater than $\overline{G}_{\mathrm{CB}}$ with $\Delta\overline{G}_{\mathrm{ii'-CB}} = 9.7$. Furthermore, the energy needed for the droplet to continue wetting the biaxial structure beyond State(ii′) is larger than the energy at State(i). As an evidence, we identify a follow-up state before reaching the underlying substrate, denoted as State(ii″), with a higher free energy than State(i), which is discussed below.

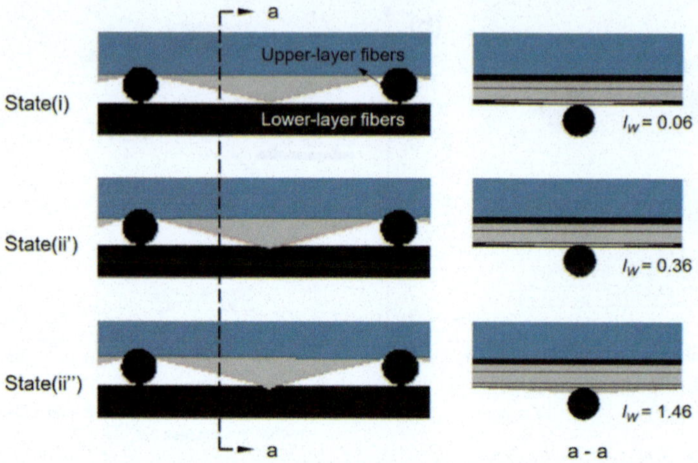

Fig. 5.8 Modeling results from Surface Evolver showing three wetting states with cross-section views. $\bar{l}_{\mathrm{w}}$ is the normalized wetting length between liquid and the lower-layer fiber

First, we imposed a constraint on Surface Evolver to limit the normalized wetting length ($\bar{l}_{\mathrm{w}}$) of each contact point between the liquid and lower-layer fibers to a small value of<0.1 to estimate $\overline{G}_{\mathrm{i}}$. Then we set $\bar{l}_{\mathrm{w}}$ to be longer than that of State(ii′) to estimate the free energy of a state that comes after State(ii′), i.e., State(ii″). Figure 5.8 shows the modeling results of State(i), State(ii′), and State (ii″) with corresponding cross-section view and $\bar{l}_{\mathrm{w}}$. The free energy of State(ii″) is already higher than that of State(i) with the normalized $\Delta\overline{G}_{\mathrm{ii'-i}} = 7.8$, even though the lower-layer fiber is not yet fully wetted at State(ii″).

The energy profile in Fig. 5.7b for the biaxial structure can explain the unexpected phenomenon in Fig. 5.4 for inter-fiber distance between 18.2 and 34 μm, that is even if the droplet touches the lower-layer fibers of a biaxial structure, the droplet could "restore" CB state. To highlight this special phenomenon, we denote this Cassie-Baxter- "restoring" process as CaRe wetting. This CaRe wetting occurs on the biaxial structure because (1) the overall greater energy barrier to the Wenzel state compared to that for the uniaxial structure hinders the droplet from further wetting the lower-layer fibers, and (2) the local minimum $\overline{G}_{\mathrm{ii'}}$ for the metastable State(ii′) is greater than $\overline{G}_{\mathrm{CB}}$ (i.e., $\Delta\overline{G}_{\mathrm{ii'-CB}} = 9.7$) which allows the droplet to reclaim the CB state for being thermodynamically more stable. It should be noted that, the porous fiber-surface also contributes to realizing the CaRe wetting because the pores in fibers ensure that the water droplet still contacts vapor phase when the droplet touches the lower layer fibers, which elevates $\overline{G}_{\mathrm{ii}}$ and makes State(ii′) less stable than the CB state.

We further study the influence of the structure parameters on the CaRe wetting, aiming at providing guidance for structure optimization. The influence is evaluated by the free

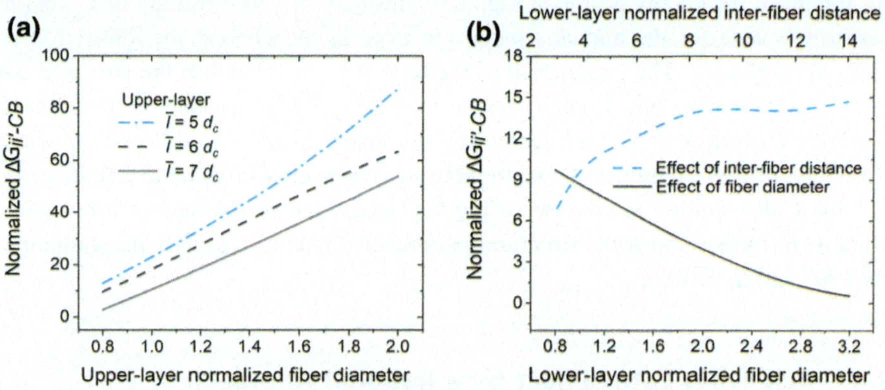

Fig. 5.9 **a** Normalized $\Delta\overline{G}_{\mathrm{ii}'-\mathrm{CB}}$ versus upper-layer fiber diameter and inter-fiber distance of the biaxial structure; the normalized low-layer fiber diameter and inter-fiber distance are 0.8 and 6, respectively. **b** Normalized $\Delta\overline{G}_{\mathrm{ii}'-\mathrm{CB}}$ versus lower-layer fiber diameter and inter-fiber distance; the normalized upper-layer fiber diameter and inter-fiber distance are 0.8 and 6, respectively

energy difference between the CB state and State(ii′), i.e., $\Delta\overline{G}_{\mathrm{ii}'-\mathrm{CB}}$; a higher $\Delta\overline{G}_{\mathrm{ii}'-\mathrm{CB}}$ facilitates the CaRe wetting and also leads to a more stable CB state. Figure 5.9 shows the influence of structure parameters on $\Delta\overline{G}_{\mathrm{ii}'-\mathrm{CB}}$.

Figure 5.9a shows a positive correlation between the increase in fiber diameter ($\overline{d}$) and the decrease in inter-fiber distance ($\overline{l}$) of the upper-layer fibers with $\Delta\overline{G}_{\mathrm{ii}'-\mathrm{CB}}$. In detail, the normalized $\Delta\overline{G}_{\mathrm{ii}'-\mathrm{CB}}$ increases 19.3 times when increasing the normalized upper-layer $\overline{d}$ from 0.8 to 2 while $\overline{l} = 7$; further decreasing $\overline{l}$ from 7 to 5 results in an additional 62.3% increase in the normalized $\Delta\overline{G}_{\mathrm{ii}'-\mathrm{CB}}$. This is because increasing $\overline{d}$ or decreasing $\overline{l}$ of the upper layer requires greater deformation and thereby more energy for the droplet to wet the lower-layer fibers.

Figure 5.9b shows the variations of $\Delta\overline{G}_{\mathrm{ii}'-\mathrm{CB}}$ with changing lower-layer $\overline{d}$ and $\overline{l}$. The normalized $\Delta\overline{G}_{\mathrm{ii}'-\mathrm{CB}}$ decreases dramatically by 94.3% when increasing the normalized lower-layer $\overline{d}$ from 0.8 to 3.2 while $\overline{l} = 6$. This is because a larger lower-layer fiber diameter means a smaller curvature of the liquid-fiber contact interface, enabling the droplet to wet the lower-layer fibers with smaller deformation; as a result, the $\overline{G}_{\mathrm{ii}'}$ is reduced. Moreover, the normalized $\Delta\overline{G}_{\mathrm{ii}'-\mathrm{CB}}$ increases by 51.0% when increasing the normalized lower-layer $\overline{l}$ from 3 to 8 while $\overline{d} = 0.8$; but further increasing $\overline{l}$ from 8 to 14 leads to a small $\Delta\overline{G}_{\mathrm{ii}'-\mathrm{CB}}$ variation of less than 4.0%. The change in $\Delta\overline{G}_{\mathrm{ii}'-\mathrm{CB}}$ with lower-layer $\overline{l}$ is attributed to the interference of liquid-fiber interfaces among adjacent lower-layer fibers caused by surface tension of liquid. For the lower-layer $\overline{l}$in the range of 3–8, the interference between adjacent liquid-fiber interfaces is reduced with increasing $\overline{l}$, resulting in a more stable CB state as reflected by the increase in $\Delta\overline{G}_{\mathrm{ii}'-\mathrm{CB}}$. However, for lower-layer $\overline{l}$ greater than 8, the interference is further reduced, leading to an insignificant effect of $\overline{l}$ on $\Delta\overline{G}_{\mathrm{ii}'-\mathrm{CB}}$.

In summary, the energy profile in Fig. 5.7b unveiled the CaRe wetting that explains the observed CB state for the biaxial structure in experiment whereas the uniaxial structure reaches Wenzel state. The occurrence of CaRe wetting is related to the structure parameters of the biaxially aligned porous fibers: increasing the upper-layer $\overline{d}$ (from 0.8 to 2), lower-layer $\overline{l}$ (from 3 to 8), and decreasing the upper-layer $\overline{l}$ (from 7 to 5), lower-layer $\overline{d}$ (from 3.2 to 0.8) facilitates the CaRe wetting and results in a stable CB state. In this study, the CaRe wetting is considered the key advantage of the biaxial structure which enables us to further adjust the structural parameters to achieve superhydrophobicity (i.e., $\theta_A > 150°$ and $\alpha_r < 10°$).

5.2.4 Achieving Lotus Effect by a Rational Approach

This section follows the guidance obtained from Fig. 5.9 to construct the biaxial structure that facilitates CaRe wetting and further adjusts the structure parameters by experiments to realize superhydrophobicity. Considering the practical fabrication range of electrospinning and the preceding guidance, we construct the upper-layer fibers with diameter of 8.1 ± 0.62 μm, while the lower-layer fibers are smaller in diameter at 3.95 ± 0.25 μm. This combination of fiber diameters can promote CaRe wetting, which enables us to strategically approach superhydrophobicity by reducing the liquid-solid contact fraction f through increasing fiber pore size and inter-fiber distance of the upper layer.

Figure 5.10 shows the changes in the measured θ_A and α_r on the biaxial structure with decreasing f. The decrease of f from 0.187 to 0.101 in the yellow shaded area is attributed to the increase of the pore size from 77 ± 16 to 180 ± 49 nm in fiber-surface (while l is kept at 32.0 ± 4.2 μm). The hydrophobicity of the biaxial structure is improved with increasing pore size as evidenced by the increase of θ_A from $141.0 \pm 0.4°$ to $157.3 \pm 0.4°$ and the decrease of α_r from $48.0 \pm 0.4°$ to $28.4 \pm 0.8°$. In addition, the upper-layer l is further adjusted with the largest pore size of 180 ± 49 nm. As shown in the blue area of Fig. 5.10, reducing f from 0.101 to 0.047 by increasing the upper-layer l further enhances hydrophobicity. The fiber-dressed surface becomes superhydrophobic when f is below 0.053 (corresponding to upper-layer l of 81.5 μm). So, the adopted biaxial structure for further studies has f values between 0.053 and 0.047. Further decreasing f below 0.047 by increasing the l is not adopted because keeping the upper-layer fibers in a closer distance facilitates the CaRe wetting as shown in Fig. 5.9a.

In contrast to the earlier studies of developing electrospun fiber-based superhydrophobic structures that used a single droplet volume to measure θ_A and α_r (or a larger droplet volume to measure α_r), this study evaluates the hydrophobicity for a range of droplets between 3 and 10 μL, as shown in Fig. 5.11. The studied droplet volume of 3–10 μL is an appropriate range for characterizing superhydrophobicity by sessile drop method [40].

The results in Fig. 5.11 show that the influence of droplet volume on θ_A is negligible because θ_A maintains relatively stable between $159.3 \pm 0.9°$ and $161.5 \pm 0.7°$ with various

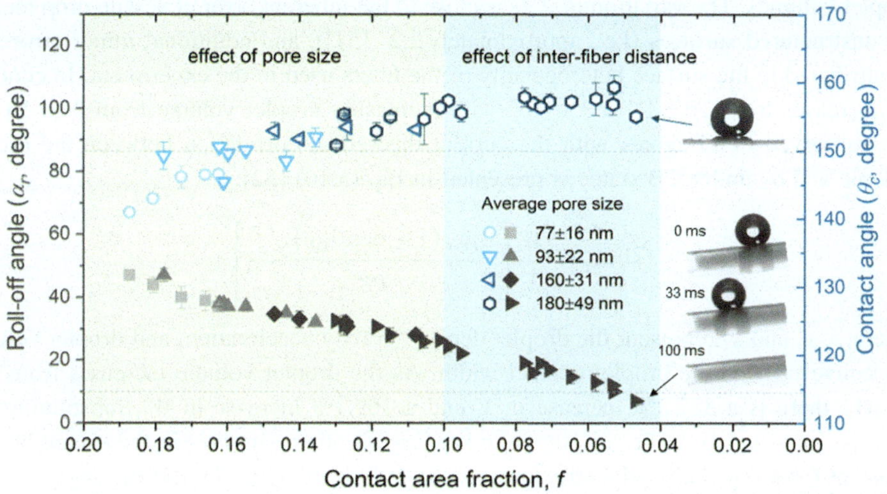

Fig. 5.10 Influence of upper-layer fiber pore size and inter-fiber distance on the measured contact angle and roll-off angle. The fiber diameters for the upper- and lower-layer are 8.1 ± 0.62 μm and 3.95 ± 0.25 μm, respectively. The inter-fiber distance for both layers in the yellow shaded area and the lower-layer in the blue shaded area are 32.0 ± 4.2 μm

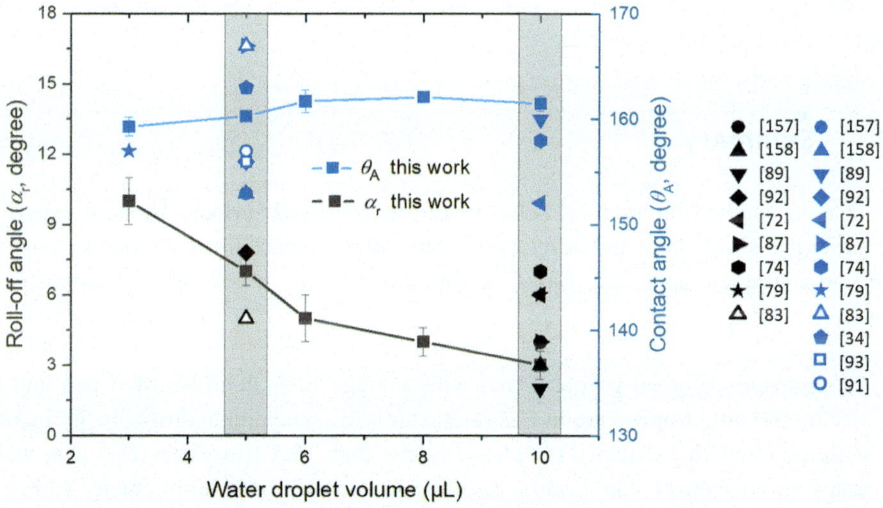

Fig. 5.11 Variation of hydrophobicity with droplet volumes on the biaxial structure, and comparison to other electrospun fiber-based superhydrophobic surfaces reported in literature [15, 18, 41–50]. The adopted biaxial structure has f between 0.053 and 0.047.

droplet volumes. The variation of 2.2° is close to the inherent error of sessile drop method for unstructured surfaces (i.e., approximately ±2° [51]), and additional minor error may be attributed to the surface heterogeneity of the fibers used in the experiment. In contrast, α_r decreases from $10 \pm 1°$ to $3 \pm 0.6°$ with increasing droplet volume from 3 to 10 μL. This observed trend agrees with the current theoretical correlation between the droplet volume and α_r under CB state, as presented in Eq. (5.10) [52]:

$$\sin \alpha_r = \frac{W}{V}\left[\frac{2\gamma_{LV}\left(1+\cos\theta_{Y,F}\right)\sqrt{f}}{\rho g}\right] \tag{5.10}$$

where ρ, g, and V represent the droplet density, gravity acceleration, and droplet volume, respectively; W is the droplet contact width. As the droplet volume increases from 3 to 10 μL, there is a 333.3% increase in V and a 169.1% increase in W. Substituting this increase of V and W in Eq. (5.10) while keeping all other variables in the square bracket constant for the given biaxial structure, results in a decrease in the roll-off angle.

It is worth noting that the superhydrophobic fibrous structure developed in this study is a one-step approach that uses only one commercially available polymer, which is deemed simple and can reduce the cost of developing electrospun superhydrophobic materials. In contrast, previous technologies typically involve complex fabrication processes and the use of multiple raw materials. For example, preparing fluorinated polymers, blending nano-size additives into solutions, and post treatment of the electrospun fibers are often required (*see* Tables 2.1 and 2.2).

5.3 Summary

This chapter studied the wetting behavior on fiber-dressed surfaces by both experiment and numerical simulation, and developed a lotus-effect surface using aligned porous fibers based on a strategic approach. Based on this research, the following conclusions can be drawn.

(1) On uniaxially aligned porous fibers with average fiber diameter of 4 μm and pore size of 160 nm, droplets are in Cassie-Baxter (CB) state and hydrophobicity increases with the inter-fiber distance (l) when l is less than 18.2 μm. After 18.2 μm, wetting transition to Wenzel state occurs, resulting in decrease of hydrophobicity with l.
(2) On biaxially aligned porous fibers with similar fiber parameters, droplets stay in CB state and hydrophobicity increases with l until $l = 34$ μm. More importantly, there is no change in trend of hydrophobicity at $l = 18.2$ μm, indicating that droplets stay in the upper-layer fibers without touching the lower-layer fibers or substrate.

(3) Energy profiles of droplets wetting the uniaxial and biaxial structures are obtained by numerical simulation in Surface Evolver. Compared to the uniaxial structure, there is a Cassie-Baxter-"restoring" (CaRe) wetting contributing to a more stable CB state in biaxial structure.
(4) The CaRe wetting ensures droplets stay in CB state even at a large l of 81.5 μm, which allows for a rational approach towards superhydrophobicity through increasing inter-fiber distance and increasing pore size. As a result, the developed structure has contact angles between 159° and 162° and roll-off angles from 10° to 3° for water droplets from 3 to 10 μL in volume.

References

1. D. Ebert, B. Bhushan, Durable Lotus-effect surfaces with hierarchical structure using micro-and nanosized hydrophobic silica particles. J. Colloid Interface Sci. **368**(1), 584–591 (2012)
2. C.-H. Chen, I.-C. Cheng, J.-Z. Chen, Facile method to convert petal effect surface to lotus effect surface for superhydrophobic polydimethylsiloxane. Surf. Interfaces **30**, 101901 (2022)
3. D. Wang, Q. Sun, M.J. Hokkanen, C. Zhang, F.-Y. Lin, Q. Liu et al., Design of robust superhydrophobic surfaces. Nature **582**(7810), 55–59 (2020)
4. R. Blossey, Self-cleaning surfaces—Virtual realities. Nat. Mater. **2**(5), 301–306 (2003)
5. B. Zhang, X. Zhao, Y. Li, B. Hou, Fabrication of durable anticorrosion superhydrophobic surfaces on aluminum substrates via a facile one-step electrodeposition approach. RSC Adv. **6**(42), 35455–35465 (2016)
6. T.V. Charpentier, A. Neville, P. Millner, R.W. Hewson, A. Morina, Development of anti-icing materials by chemical tailoring of hydrophobic textured metallic surfaces. J. Colloid Interface Sci. **394**, 539–544 (2013)
7. J.W. Gose, K. Golovin, M. Boban, J.M. Mabry, A. Tuteja, M. Perlin et al., Characterization of superhydrophobic surfaces for drag reduction in turbulent flow. J. Fluid Mech. **845**, 560–580 (2018)
8. R.N. Wenzel, Resistance of solid surfaces to wetting by water. Ind. Eng. Chem. **28**(8), 988–994 (1936)
9. A. Cassie, S. Baxter, Wettability of porous surfaces. Trans. Faraday Soc. **40**, 546–551 (1944)
10. F. Wang, S. Li, L. Wang, Fabrication of artificial super-hydrophobic lotus-leaf-like bamboo surfaces through soft lithography. Colloids Surf. A **513**, 389–395 (2017)
11. C.-H. Xue, Y.-R. Li, P. Zhang, J.-Z. Ma, S.-T. Jia, Washable and wear-resistant superhydrophobic surfaces with self-cleaning property by chemical etching of fibers and hydrophobization. ACS Appl. Mater. Interfaces **6**(13), 10153–10161 (2014)
12. K. Bachus, L. Mats, H. Choi, G. Gibson, R. Oleschuk, Fabrication of patterned superhydrophobic/hydrophilic substrates by laser micromachining for small volume deposition and droplet-based fluorescence. ACS Appl. Mater. Interfaces **9**(8), 7629–7636 (2017)
13. N. Nuraje, W.S. Khan, Y. Lei, M. Ceylan, R. Asmatulu, Superhydrophobic electrospun nanofibers. J. Mater. Chem. A **1**(6), 1929–1946 (2013)
14. M. Ma, R.M. Hill, J.L. Lowery, S.V. Fridrikh, G.C. Rutledge, Electrospun poly (styrene-block-dimethylsiloxane) block copolymer fibers exhibiting superhydrophobicity. Langmuir **21**(12), 5549–5554 (2005)

15. D.-E. Lee, E.-Y. Choi, H.-J. Yang, A.S. Murthy, T. Singh, J.-M. Lim et al., Highly stretchable superhydrophobic surface by silica nanoparticle embedded electrospun fibrous mat. J. Colloid Interface Sci. **555**, 532–540 (2019)
16. J. Lin, Y. Cai, X. Wang, B. Ding, J. Yu, M. Wang, Fabrication of biomimetic superhydrophobic surfaces inspired by lotus leaf and silver ragwort leaf. Nanoscale **3**(3), 1258–1262 (2011)
17. R. Asmatulu, M. Ceylan, N. Nuraje, Study of superhydrophobic electrospun nanocomposite fibers for energy systems. Langmuir **27**(2), 504–507 (2011)
18. M. Ma, M. Gupta, Z. Li, L. Zhai, K.K. Gleason, R.E. Cohen et al., Decorated electrospun fibers exhibiting superhydrophobicity. Adv. Mater. **19**(2), 255–259 (2007)
19. M. Ma, Y. Mao, M. Gupta, K.K. Gleason, G.C. Rutledge, Superhydrophobic fabrics produced by electrospinning and chemical vapor deposition. Macromolecules **38**(23), 9742–9748 (2005)
20. F. Zareei Pour, M.M. Sabzehmeidani, H. Karimi, V. Madadi Avargani, M. Ghaedi, Superhydrophobic–superoleophilic electrospun nanofibrous membrane modified by the chemical vapor deposition of dimethyl dichlorosilane for efficient oil–water separation. J. Appl. Polym. Sci. **136**(24), 47621 (2019)
21. S. Jin, Y. Park, C.H. Park, Preparation of breathable and superhydrophobic polyurethane electrospun webs with silica nanoparticles. Text. Res. J. **86**(17), 1816–1827 (2016)
22. A. Rawal, Design parameters for a robust superhydrophobic electrospun nonwoven mat. Langmuir **28**(6), 3285–3289 (2012)
23. B. Emami, H.V. Tafreshi, M. Gad-el-Hak, G. Tepper, Effect of fiber orientation on shape and stability of air–water interface on submerged superhydrophobic electrospun thin coatings. J. Appl. Phys. **111**(6), 064325 (2012)
24. T. Onda, Theoretical investigation of wenzel and cassie wetting states on porous films and fiber meshes. Langmuir **38**(45), 13744–13752 (2022)
25. Y. Xue, S. Chu, P. Lv, H. Duan, Importance of hierarchical structures in wetting stability on submersed superhydrophobic surfaces. Langmuir **28**(25), 9440–9450 (2012)
26. C. Huang, N.L. Thomas, Fabrication of porous fibers via electrospinning: strategies and applications. Polym. Rev. **60**(4), 595–647 (2020)
27. J. Berthier, K.A. Brakke, *The Physics of Microdroplets* (Wiley, 2012)
28. D.L. Hu, B. Chan, J.W. Bush, The hydrodynamics of water strider locomotion. Nature **424**(6949), 663–666 (2003)
29. S. Brandon, N. Haimovich, E. Yeger, A. Marmur, Partial wetting of chemically patterned surfaces: the effect of drop size. J. Colloid Interface Sci. **263**(1), 237–243 (2003)
30. L. He, W. Liang, Z. Wang, A. Akbarzadeh, A three-dimensional model for analyzing the anisotropic wetting behavior of striped surfaces. Colloids Surf. A **552**, 67–74 (2018)
31. H. Fashandi, M. Karimi, Pore formation in polystyrene fiber by superimposing temperature and relative humidity of electrospinning atmosphere. Polymer **53**(25), 5832–5849 (2012)
32. Y. Miyauchi, B. Ding, S. Shiratori, Fabrication of a silver-ragwort-leaf-like super-hydrophobic micro/nanoporous fibrous mat surface by electrospinning. Nanotechnology **17**(20), 5151 (2006)
33. L. Jiang, Y. Zhao, J. Zhai, A lotus-leaf-like superhydrophobic surface: a porous microsphere/nanofiber composite film prepared by electrohydrodynamics. Angew. Chem.-Int. Ed. **43**(33), 4338–4341 (2004)
34. P. Tsai, R.G. Lammertink, M. Wessling, D. Lohse, Evaporation-triggered wetting transition for water droplets upon hydrophobic microstructures. Phys. Rev. Lett. **104**(11), 116102 (2010)
35. F. Schellenberger, N. Encinas, D. Vollmer, H.-J. Butt, How water advances on superhydrophobic surfaces. Phys. Rev. Lett. **116**(9), 096101 (2016)
36. E. Bormashenko, Progress in understanding wetting transitions on rough surfaces. Adv. Coll. Interface. Sci. **222**, 92–103 (2015)

37. A. Marmur, The Lotus effect: superhydrophobicity and metastability. Langmuir **20**(9), 3517–3519 (2004)
38. A. Dupuis, J.M. Yeomans, Modeling droplets on superhydrophobic surfaces: equilibrium states and transitions. Langmuir **21**(6), 2624–2629 (2005)
39. Y. Jiang, J. Lian, Z. Jiang, Y. Li, C. Wen, Thermodynamic analysis on wetting states and wetting state transitions of rough surfaces. Adv. Colloid Interface Sci. **278**, 102136 (2020)
40. C.E. Cansoy, The effect of drop size on contact angle measurements of superhydrophobic surfaces. RSC Adv. **4**(3), 1197–1203 (2014)
41. H. Gu, G. Li, P. Li, H. Liu, T.T. Chadyagondo, N. Li et al., Superhydrophobic and breathable SiO2/polyurethane porous membrane for durable water repellent application and oil-water separation. Appl. Surf. Sci. **512**, 144837 (2020)
42. D. Han, A.J. Steckl, Superhydrophobic and oleophobic fibers by coaxial electrospinning. Langmuir **25**(16), 9454–9462 (2009)
43. L. Zhou, C.L. Li, P.T. Chang, S.H. Tan, A.L. Ahmad, S.C. Low, Intrinsic microspheres structure of electrospun nanofibrous membrane with rational superhydrophobicity for desalination via membrane distillation. Desalination **527**, 115594 (2022)
44. W.S. Wong, P. Gutruf, S. Sriram, M. Bhaskaran, Z. Wang, A. Tricoli, Strain engineering of wave-like nanofibers for dynamically switchable adhesive/repulsive surfaces. Adv. Func. Mater. **26**(3), 399–407 (2016)
45. K. Acatay, E. Simsek, C. Ow-Yang, Y.Z. Menceloglu, Tunable, superhydrophobically stable polymeric surfaces by electrospinning. Angew. Chem. **116**(39), 5322–5325 (2004)
46. J.-M. Lim, G.-R. Yi, J.H. Moon, C.-J. Heo, S.-M. Yang, Superhydrophobic films of electrospun fibers with multiple-scale surface morphology. Langmuir **23**(15), 7981–7989 (2007)
47. T.-T. Li, Y. Wang, H.-K. Peng, X. Zhang, B.-C. Shiu, J.-H. Lin et al., Lightweight, flexible and superhydrophobic composite nanofiber films inspired by nacre for highly electromagnetic interference shielding. Compos. A Appl. Sci. Manuf. **128**, 105685 (2020)
48. J. Sheng, Y. Xu, J. Yu, B. Ding, Robust fluorine-free superhydrophobic amino-silicone oil/SiO2 modification of electrospun polyacrylonitrile membranes for waterproof-breathable application. ACS Appl. Mater. Interfaces **9**(17), 15139–15147 (2017)
49. X. Yu, Y. Li, X. Wang, Y. Si, J. Yu, B. Ding, Thermoconductive, moisture-permeable, and superhydrophobic nanofibrous membranes with interpenetrated boron nitride network for personal cooling fabrics. ACS Appl. Mater. Interfaces **12**(28), 32078–32089 (2020)
50. X. Yu, Y. Li, X. Yin, X. Wang, Y. Han, Y. Si et al., Corncoblike, superhydrophobic, and phase-changeable nanofibers for intelligent thermoregulating and water-repellent fabrics. ACS Appl. Mater. Interfaces **11**(42), 39324–39333 (2019)
51. Y. Yuan, T.R. Lee, Contact angle and wetting properties, in *Surface Science Techniques* (2013), pp. 3–34
52. C. Lv, C. Yang, P. Hao, F. He, Q. Zheng, Sliding of water droplets on microstructured hydrophobic surfaces. Langmuir **26**(11), 8704–8708 (2010)

6 Potential Applications of the "Lotus" Surface Based on Biaxially Aligned Porous Fibers

Electrospun fibrous structures are featured with high porosity, up to more than 90% [1], with interconnectivity between the inter-fiber pores, which can realize a high breathability. Thus, superhydrophobic and breathable structure, which has wide engineering applications such as wearable electronics [2], personal protections [3], gas sensors [4], etc., can be developed by electrospinning.

The biaxial structure with lotus effect developed in Chap. 5 only consists of two layers of fibers, realizing a total thickness less than 20 μm. In contrast, most earlier reported lotus-effect electrospun structures consist of randomly distributed fibers, so usually a thicker (more than 50 μm [5, 6]) fibrous mat is made to prevent droplet from touching the underneath substrate, which results in Wenzel state. However, increasing structure thickness reduces breathability as air and water vapor needs to transmit the structure with greater resistance. So, it is expected that the biaxial structure can have a high breathability (to be measured in the following section).

In this chapter, two potential applications of the biaxial structure are demonstrated based on the lotus superhydrophobicity and high breathability: developing waterproof, breathable and superhydrophobic membrane, and as a surface layer for protecting wearable electronics.

6.1 Waterproof, Breathable and Superhydrophobic Membrane

The developed biaxial structure of aligned porous fibers in Chap. 5 offers an opportunity to address the challenge of developing a simple method for producing waterproof, breathable, and superhydrophobic membrane. The waterproof and breathable membrane

Y. Zhang, *Superhydrophobicity of Surfaces Dressed by Electrospun Fibers*, Synthesis Lectures on Green Energy and Technology, https://doi.org/10.1007/978-3-031-55552-7_6

that prevents liquid water permeation while allows moisture and air transmission are highly desired for various applications, such as wearable electronics, medical dressing, and desalination [7]. Lotus superhydrophobicity is an additional desired feature for waterproof and breathable membranes because it enables self-cleaning, which help to maintain the waterproofness and breathability. For fabrication, electrospinning is widely adopted for developing waterproof breathable membranes because the electrospun fibrous structure provides a good basis for breathability [8, 9] owing to its interconnected pores. However, despite the substantial progress made in this field, existing technologies can only achieve two or three out of four features of waterproofing, breathability, superhydrophobicity, and simplicity (see Table 6.1). The demonstration of our approach that achieves all four features is shown as follows.

Figure 6.1 demonstrates the application of biaxially aligned porous fibers for developing water-proof, breathable, and superhydrophobic (WBS) membrane. The fibers are dressed on a commercial nylon mesh that is originally water permeable. Despite the biaxially aligned porous fibers being sparsely arranged with only two layers of fibers, they transform the mesh into a waterproof membrane. As shown in Fig. 6.1a, the water passes the pristine mesh and enters the bottom vessel, while the fiber-dressed mesh is capable of holding the water in the top vessel.

Figure 6.1b indicates the breathability of the fiber-dressed mesh. Specifically, water vapor generated from heating passes through the fiber-dressed mesh and condenses in the upper vessel. For quantification, the breathability is assessed by water vapor transmission rate (WVTR) test following the ASTM E96 standard. In a typical test, $CaCl_2$ desiccant is sealed by the tested sample in a testing cup with a 25 cm^2 opening, which is placed in an environmental chamber controlled at 38 °C and 90% RH. The weight of the testing cup is measured every 10 min to calculate the WVTR using Eq. (6.1).

$$\mathrm{WVTR} = \frac{\Delta m}{\Delta t \times A_{\mathrm{c}}} \tag{6.1}$$

where Δm is the weight change of $CaCl_2$ in Δt period and A_{c} is the cup opening area.

The WVTR of the fiber-dressed mesh is measured to be 20.8 ± 0.1 kg m^{-2} d^{-1}. To the best of our knowledge, this WVTR is the highest value reported so far for electrospun WBS membranes (see Table 6.1) due to its large inter-fiber distance and low thickness. Moreover, the biaxially aligned porous fibers also impart superhydrophobicity to the mesh, as indicated by the rapid rolling of a water droplet from its surface in Fig. 6.1c.

It should be noted that the nylon mesh substrate herein is only for demonstration. The developed biaxial structure is essentially a versatile dressing layer that can be applied to various substrates, such as conventional electrospun membranes, microporous media, metal mesh, etc. The choice of substrate depends on mechanical strength, hydrostatic pressure, breathability, and so on, in various applications. For example, Fig. 6.2 demonstrates the modification of a pristine electrospun PVDF-HFP membrane. Prior to modification, water droplets are observed to be pinned on the surface of the PVDF-HFP membrane,

Table 6.1 Comparison of electrospun waterproof, breathable and superhydrophobic membrane

Materials	Fabrication	W[1]	B[2]	S[3]	SP[4]	WVTR (kg m^2 d^{-1})	WVTR testing condition	Refs.
FPU/BN	Blending fluorinated polymer and nanosheets for electrospinning	✓	✓	✓		11.6	38 °C, 50% RH[5]	[10]
FPU/PU/PCC	Blending fluorinated polymer and nano-capsules for electrospinning	✓	✓	✓		11.4	38 °C, 50% RH[5]	[11]
PU	Electrospinning of nanosized fibers	✓	✓		✓	1.5	25 °C, 65% RH[5]	[12]
Silk fibroin	Electrospinning of nanosized fibers		✓		✓	2.0±0.1	38 °C, 50% RH[5]	[13]
PU/SiO_2	Post-treatments of surface grafting and hydrothermal reaction to electrospun fibers for growing nanoparticles	✓	✓	✓		8.4	–	[5]
PVDF	Post-treatments of thermo-pressing to electrospun fibers	✓	✓			10.9	38 °C, 90% RH[6]	[14]
PVDF/PVB	Blending two polymers for electrospinning and followed by thermal treatment	✓	✓			10.6	38 °C, 90% RH[6]	[15]

(continued)

Table 6.1 (continued)

Materials	Fabrication	W[1]	B[2]	S[3]	SP[4]	WVTR ($kg\ m^2\ d^{-1}$)	WVTR testing condition	Refs.
PVDF/PVB/ octadecane/ CNT	Blending nanosize carbon tubes with polymer solutions for co-axial electrospinning	✓	✓			7.8	38 °C, 90% RH[6]	[16]
PAN/FPU	Post thermal treatment to electrospun fibers	✓	✓			10.1	38 °C, 90% RH[6]	[17]
PES/$BaTiO_3$	Blending nanoparticles and polymer solution for electrospinning		✓			6.2	– [6]	[18]
PS	Electrospinning of aligned porous fibers	✓	✓	✓	✓	20.8 ± 0.1	38 °C, 90% RH[6]	This work

[1] Waterproof; [2] Breathable; [3] Superhydrophobic; [4] Simple; [5] Test method based on water evaporation; [6] Test method based on vapor absorption by desiccant

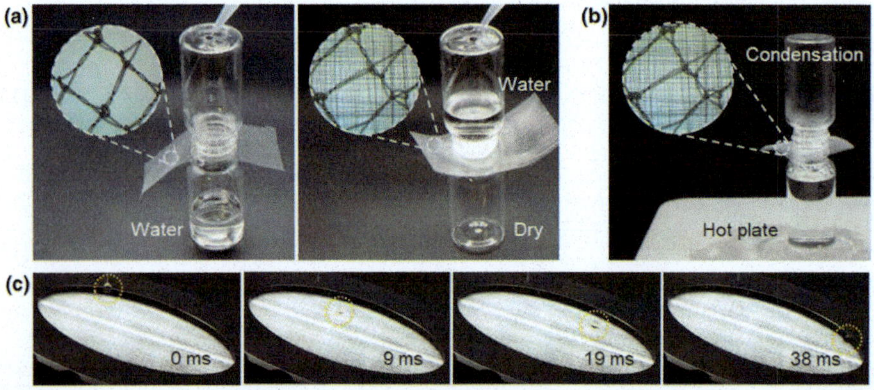

Fig. 6.1 Development of waterproof, breathable and superhydrophobic membrane by applying biaxially aligned porous fibers on a mesh. Demonstration for **a** waterproofing, **b** breathability, and **c** superhydrophobicity of the fiber-dressed mesh. Insets of **a**, **b** are the optical microscopic images of the mesh before and after being dressed by the fibers

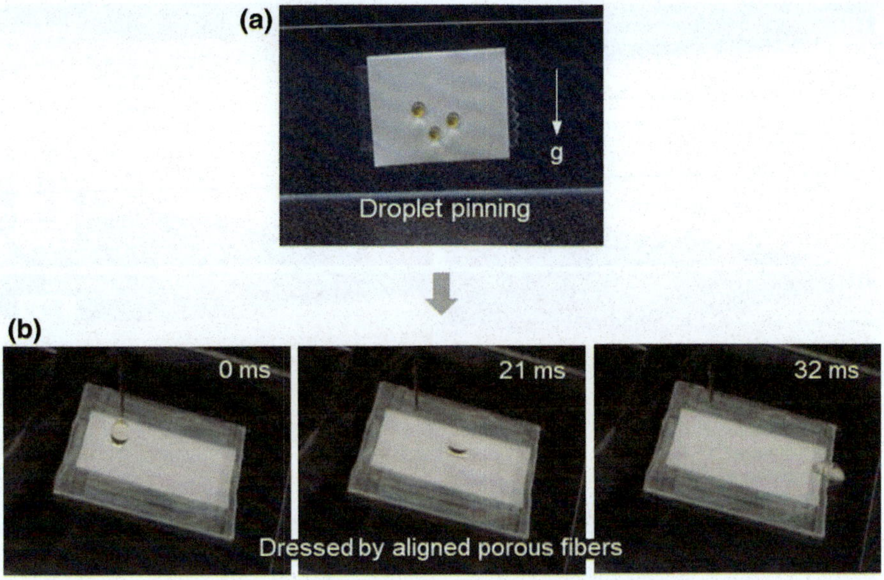

Fig. 6.2 Modifying an electrospun PVDF-HFP membrane from **a** droplet-pinning to **b** lotus superhydrophobicity by dressing the biaxially aligned porous fibers

even when placing the membrane vertically as shown in Fig. 6.2a. In contrast, after dressing the membrane with the biaxially aligned porous fibers, the droplet rapidly rolls off the surface as shown in Fig. 6.2b.

6.2 Surface Layer for Wearable Electronics

The emerging wearable electronics have advanced considerably for healthcare monitoring, energy harvesting, human-machine interface, and so on. To ensure user comfort and device longevity, it is important to incorporate a breathable superhydrophobic surface layer into the wearable electronics. The surface layer allows for sweat evaporation, preventing skin irritation and inflammation from prolonged use [19], and protects the underlying electronics from wetting damage. This section demonstrates such an application, where a flexible nylon web, representing an electronic layer, is patched onto a human forearm with a surface layer of the developed biaxially aligned porous fibers.

Figure 6.3a, b show the temperature change of the skin being patched by a non-breathable adhesive tape and the developed biaxial structure, respectively. The temperature change of the skin after being covered for 60 min reflects the influence of the patched material on sweat evaporation. A material with inadequate breathability causes a temperature drop due to the evaporation of the accumulated sweat, see Fig. 6.3a. In contrast, the

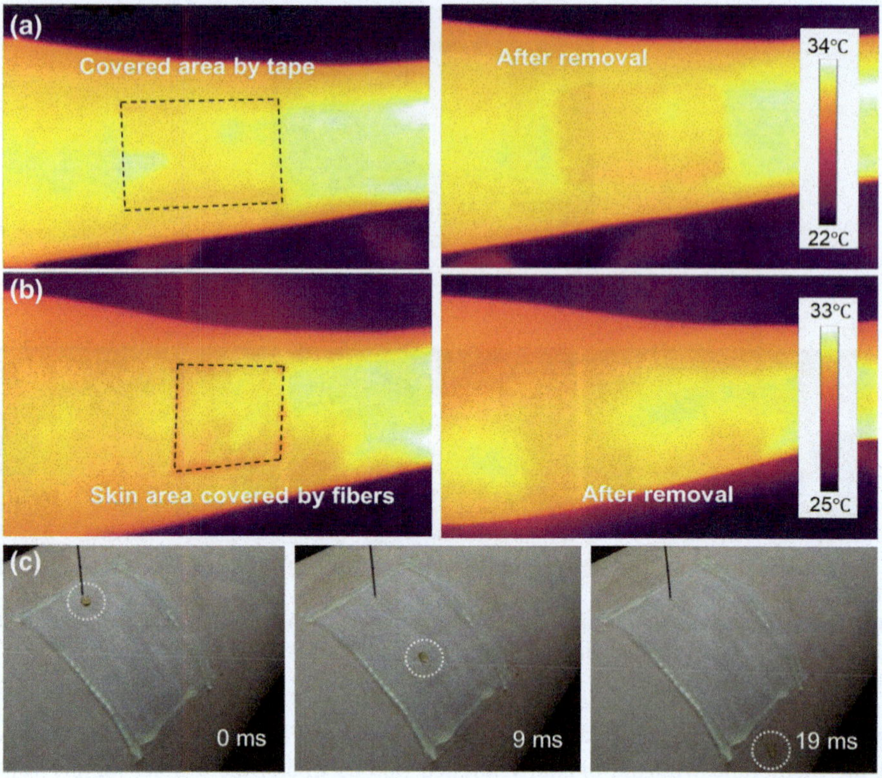

Fig. 6.3 Development of the surface layer for protecting wearable electronics. Thermal camera images of human forearm skin, which is patched by **a** a non-breathable adhesive tape and **b** the developed biaxial structure for 60 min and removal afterward. **c** Superhydrophobicity of forearm skin area covered by the developed biaxial structure. Water droplet is dyed in orange

skin shows unnoticeable temperature change after being covered by the biaxial structure as shown in Fig. 6.3b, indicating sufficient breathability for wearable uses. In addition, Fig. 6.3c indicates that a water droplet can rapidly roll off the same biaxial structure, which demonstrates its water repellence if used as a surface layer for wearable electronics on human skin. This characteristic is particularly useful in protecting electronics devices that are sensitive to water intrusion [20], such as wearable nano-generators.

In addition, Fig. 6.4 shows that the developed biaxial structure has a high transparency as the pattern under the fiber-dressed glass slide is visible. The transparency of the biaxial structure is characterized using a UV-Vis spectrophotometer. The average transmittance in the visible region (380–760 nm) is $83.6 \pm 3.1\%$. The transparency can be useful for allowing direct observation and close monitoring of wearable electronics.

Fig. 6.4 Transparency of the developed biaxially aligned porous fibers

References

1. Z. Yan, J. Xiong, B. Wang, M. Gao, G. Yin, T. Hu et al., Recent advances in breathable electronics. Nano Res. 1–13 (2022)
2. J. Luo, S. Gao, H. Luo, L. Wang, X. Huang, Z. Guo et al., Superhydrophobic and breathable smart MXene-based textile for multifunctional wearable sensing electronics. Chem. Eng. J. **406**, 126898 (2021)
3. L. Lao, D. Shou, Y. Wu, J. Fan, "Skin-like" fabric for personal moisture management. Sci. Adv. **6**(14), eaaz0013 (2020)
4. X. Zhang, X. Hao, Z. Zhai, J. Wang, H. Li, Y. Sun et al., Flexible H2S sensors: Fabricated by growing NO2-UiO-66 on electrospun nanofibers for detecting ultralow concentration H2S. Appl. Surf. Sci. **573**, 151446 (2022)
5. H. Gu, G. Li, P. Li, H. Liu, T.T. Chadyagondo, N. Li et al., Superhydrophobic and breathable SiO2/polyurethane porous membrane for durable water repellent application and oil-water separation. Appl. Surf. Sci. **512**, 144837 (2020)
6. L. Zhou, C.L. Li, P.T. Chang, S.H. Tan, A.L. Ahmad, S.C. Low, Intrinsic microspheres structure of electrospun nanofibrous membrane with rational superhydrophobicity for desalination via membrane distillation. Desalination **527**, 115594 (2022)
7. D. Ebert, B. Bhushan, Wear-resistant rose petal-effect surfaces with superhydrophobicity and high droplet adhesion using hydrophobic and hydrophilic nanoparticles. J. Colloid Interface Sci. **384**(1), 182–188 (2012)
8. A. Marmur, Wetting on hydrophobic rough surfaces: to be heterogeneous or not to be? Langmuir **19**(20), 8343–8348 (2003)
9. C. H. Kung, P.K. Sow, B. Zahiri, W. Mérida, Assessment and interpretation of surface wettability based on sessile droplet contact angle measurement: challenges and opportunities. Adv. Mater. Interfaces **6**(18), 1900839 (2019)
10. X. Yu, Y. Li, X. Wang, Y. Si, J. Yu, B. Ding, Thermoconductive, moisture-permeable, and superhydrophobic nanofibrous membranes with interpenetrated boron nitride network for personal cooling fabrics. ACS Appl. Mater. Interfaces **12**(28), 32078–32089 (2020)
11. X. Yu, Y. Li, X. Yin, X. Wang, Y. Han, Y. Si et al., Corncoblike, superhydrophobic, and phase-changeable nanofibers for intelligent thermoregulating and water-repellent fabrics. ACS Appl. Mater. Interfaces **11**(42), 39324–39333 (2019)
12. F.J. Maksoud, M. Lameh, S. Fayyad, N. Ismail, A.R. Tehrani-Bagha, N. Ghaddar et al., Electrospun waterproof breathable membrane with a high level of aerosol filtration. J. Appl. Polym. Sci. **135**(2), 45660 (2018)

13. S. Ghalei, J. Nourmohammadi, A. Solouk, H. Mirzadeh, Enhanced cellular response elicited by addition of amniotic fluid to alginate hydrogel-electrospun silk fibroin fibers for potential wound dressing application. Colloids Surf. B Biointerfaces **172**, 82–89 (2018)
14. X. Li, J. Lin, F. Bian, Y. Zeng, Improving waterproof/breathable performance of electrospun poly(vinylidene fluoride) fibrous membranes by thermo-pressing. J. Polym. Sci. Part B: Polym. Phys. **56**(1), 36–45 (2018)
15. M. Zhang, J. Sheng, X. Yin, J. Yu, B. Ding, Polyvinyl butyral modified polyvinylidene fluoride breathable-waterproof nanofibrous membranes with enhanced mechanical performance. Macromol. Mater. Eng. **302**(8) (2017)
16. L. Yi, S. Wang, L. Wang, J. Yao, J. Marek, M. Zhang, A waterproof and breathable nanofibrous membrane with thermal-regulated property for multifunctional textile application. J. Appl. Polym. Sci. **138**(19), 50391 (2020)
17. J. Sheng, Y. Li, X. Wang, Y. Si, J. Yu, B. Ding, Thermal inter-fiber adhesion of the polyacrylonitrile/fluorinated polyurethane nanofibrous membranes with enhanced waterproof-breathable performance. Sep. Purif. Technol. **158**, 53–61 (2016)
18. N. Wang, M. Cai, X. Yang, Y. Yang, Electret nanofibrous membrane with enhanced filtration performance and wearing comfortability for face mask. J. Colloid Interface Sci. **530**, 695–703 (2018)
19. Y. Zhu, R. Haghniaz, M.C. Hartel, S. Guan, J. Bahari, Z. Li et al., A breathable, passive-cooling, non-inflammatory, and biodegradable aerogel electronics for wearable physical-electrophysiological-chemical analysis. Adv. Mater. 2209300 (2022)
20. X. Gong, X. Yin, F. Wang, X. Liu, J. Yu, S. Zhang et al., Electrospun nanofibrous membranes: a versatile medium for waterproof and breathable application. Small **19**(2), 2205067 (2023)

Concluding Insights 7

7.1 Conclusions

Based on the wetting behavior on fiber-dressed surfaces, this book develops rose-petal and lotus superhydrophobic surfaces using uniaxially aligned fibers and biaxially aligned porous fibers, respectively, both fabricated by simple one-step electrospinning.

An integrated thermodynamic model for droplet wetting on various parallel-structured surfaces is established to understand the wetting behavior. The model relates free energy with the continuous movement of the three-phase contact line (TPCL) in different wetting states, and the forming conditions of energy barrier are derived from the model. Then, the model are validated using existing works and the experimental data by aligned fibers. The model indicates that for surfaces patterned by sharp-edge structures (cross-section shape) with an internal angle φ and an intrinsic contact angle θ_Y, the initiating and ending apparent contact angles (APCA, θ_A) for energy barrier in noncomposite state are $\theta_Y + \varphi$ and $\theta_Y - \varphi$, respectively. For the round structures, an energy barrier may form for every structure at $\theta_A + (180^\circ - \theta_Y)$ in noncomposite state because the round surface has continuous inclination angles with respect to the horizontal plane. Thus, surfaces patterned with round structures, such as fibers, can achieve a large contact angle with a high contact angle hysteresis, which is needed for petal superhydrophobicity.

Then this book studies the influence of fiber parameters on the surface wettability and develops petal-effect surfaces based on uniaxially aligned fibers. First, the wetting model for surfaces dressed by aligned fibers is specified based on the integrated model mentioned above. The model indicates that the magnitude of energy barrier, which can pin a droplet at a higher APCA than the equilibrium contact angle, increases with the fiber diameter and the basal width of droplet. Uniaxially aligned polystyrene fibers with average diameters of 0.9 and 1.8 μm and varying inter-fiber distance (l) from 70 to 1450 μm are fabricated by one-step electrospinning. At large l > 800 μm, where two fibers can

Y. Zhang, *Superhydrophobicity of Surfaces Dressed by Electrospun Fibers*, Synthesis Lectures on Green Energy and Technology, https://doi.org/10.1007/978-3-031-55552-7_7

pin a 2 μL droplet, the maximum APCA of the 1.8 μm fibers is 155°, compared to 147° for 0.9 μm fibers. The greater APCA is due to the higher magnitude of energy barrier for 1.8 μm fibers. When l is below 800 μm, two 1.8 μm fibers are unable to pin the droplet because of the reduced energy barrier with basal width, so three fibers are wetted. Similar variation pattern of APCA by reducing l are observed for three to five fibers. However, there is an increase of the overall APCAs due to that more energy barriers are induced to attenuate the perturbation energy.

Petal-effect surfaces are developed based on uniaxially aligned polystyrene fibers with diameter of 1.8 μm and inter-fiber distance between 70 and 130 μm. The APCAs of the fiber-dressed surfaces are in a range of 150°–156°, and droplets can adhere to the surfaces even by turning surfaces upside down.

After that, this book studies the wetting behaviors on surfaces dressed by aligned porous fibers and develops a lotus-effect surface using biaxially aligned porous fibers. In experiment, on uniaxially aligned porous fibers with average fiber diameter of 4 μm and pore size of 160 nm, droplets are in Cassie-Baxter (CB) state and hydrophobicity increases with l when l is below 18.2 μm. After 18.2 μm, the wetting transition to Wenzel state occurs, resulting in decrease of hydrophobicity with l. On biaxially aligned porous fibers with similar fiber parameters, droplets stay in CB state and hydrophobicity increases with l until $l = 34$ μm. More importantly, there is no change in the trend of hydrophobicity at $l = 18.2$ μm until 34 μm, indicating that droplets stay in the upper-layer fibers without touching the lower-layer fibers or substrate. To explain the unchanged trend, the energy profiles of droplets wetting uniaxial and biaxial structures are obtained using numerical simulation in Surface Evolver. Compared to the uniaxial structure, a Cassie-Baxter-"restoring" (CaRe) wetting results in a more stable CB state in biaxial structure.

Lotus-effect surfaces are developed based on the CaRe wetting of biaxially aligned porous fibers. The CaRe wetting allows for a rational approach towards superhydrophobicity by increasing inter-fiber distance to 81.5 μm and increasing pore sizes to 180 nm. As a result, the developed structure display contact angles between 159° and 162° and roll-off angles from 10° to 3° for water droplets from 3 to 10 μL in volume.

Finally, the biaxially aligned porous fibers are used to develop a waterproof, breathable, and superhydrophobic membrane by depositing the fibers on a supporting nylon mesh. The biaxial structure transforms the nylon mesh into a waterproof membrane. The breathability of the membrane is characterized by water vapor transmission rate (WVTR) test, resulting in a WVTR of 20.8 ± 0.1 kg m^{-2} d^{-1}. The membrane is also shown to be superhydrophobic as water droplets rapidly roll off the surface. The second application is to develop a surface layer for protecting wearable electronics. The skin of a human forearm shows unnoticeable temperature change after being patched with a surface layer of the biaxial structure for 60 min. Moreover, the average transmittance in visible region (380–760 nm) of the biaxial structure is characterized to be $83.6 \pm 3.1\%$ using a UV-Vis spectrophotometer.

Overall, this book provides insights into the interfacial behavior of wetting fiber-dressed surfaces, and a rational design of surfaces with controlled wettability, especially rose-petal and lotus superhydrophobicity, using electrospun fibers.

7.2 Recommended Future Works

Based on this book, the following recommendations are made for future research works.

(1) **Improving the alignment and distribution uniformity of fibers**. The aligned fibers used in this book are fabricated using a conventional electrospinning setup with a rotating fin collector. However, the alignment and uniformity of distribution of fibers are hard to be precisely controlled, as shown in Fig. 5.2e, f, because of the bending instabilities of the jets in electrospinning [1].

Improving the alignment and uniformity of distribution helps to more efficiently control the surface wettability. For example, to develop lotus-effect surfaces using aligned fibers, a higher distribution density of fibers is needed to prevent droplets from penetration if the fibers are poorly aligned and distributed with low uniformity. However, the high density can increase liquid-fiber contact area, causing decreasing of hydrophobicity.

Even though a higher rotating speed of the fin collector can improve the alignment and distribution to some extent, extra stretching is exerted on the unsolidified jets by the collector [2], which may alter the fiber morphology. In addition to the collector rotating speed, the relative humidity, applied voltage, and spinning distance can also influence the alignment of the produced fibers.

Near-field electrospinning (NFES) may be one solution to precisely control the alignment and uniformity of distribution of fibers because it can control the deposition of fibers. The bending instabilities of the jets are avoided in NFES by reducing the spinning distance to several millimeters or below [3].

(2) **Further investigating the surface wettability of aligned fibers**. For the aligned non-porous fibers, the experimental lower limit of inter-fiber distance in this book is around 70 μm. This lower limit ensures a noncomposite wetting state for the rose-petal effect. However, further reducing the inter-fiber distance may generate intersections of adjacent fibers, which can change the magnitude of free energy barrier, because of the limited alignment of fibers by conventional electrospinning.

Studying the wetting behavior of droplets for inter-fiber distance lower than 70 μm helps to develop the full spectrum of wettability of fiber-dressed surfaces, facilitating the wetting control, especially for the region between lotus effect and petal effect, by electrospun fibers. In literature, Aziz et al. [4] have established an analytical model to determine the critical inter-fiber distance for the composite state on aligned fibers, and numerically calculated the anisotropic contact angles. Moreover,

Meng et al. [5] experimentally studied the wetting behavior on densely aligned fibers with inter-fiber distance below 30 μm. However, super-hydrophobicity is not achieved partially because of the poor alignment of fibers. Therefore, future works may modify the collector for conventional electrospinning or adopt NFES to produce well-aligned fibers with lower inter-fiber distances.

(3) **Enhancing the mechanical strength of the fibrous structure**. As shown in Chaps. 4 and 5, the structure, especially the morphology and spatial distribution, of fibers is critical in modifying surface wettability. For instance, increasing the inter-fiber distance may cause the wetting transition of droplets from Cassie-Baxter to Wenzel state, as indicated in Fig. 5.4a. Therefore, the mechanical strength of the fibrous structure can influence the efficiency and duration of wetting control. However, electrospun fibers have been suffered from a low mechanical strength [6].

To enhance the mechanical strength, future works may focus on the following methods. The first method is to use polymers with higher mechanical strength for electrospinning such as PVDF [7], which also has an intrinsic contact angle close to PS. Second, fibers can be cross-linked at intersections. The cross-linking can not only increase the overall strength of the fibrous structure, but also fix the position of each fiber. Third, for the biaxial structure, more layers of fibers can be added at the bottom since the lotus effect is created by the top two layers. However, this may reduce the breathability of the structure.

(4) **Achieving controlled motion of droplets based on the biaxial structure**. Droplet manipulation attracts much attention for its important role in fields like microfluidics [8]. The biaxial structure in this book, which enables droplets easily rolling off, can be modified to achieve controlled motion of droplets.

The modified biaxial structure also contains two layers of fibers: the lower layer still uses aligned porous fibers; the upper layer is produced using NFES to pattern fibers with various shapes, e.g., "S" in shape, but the fiber parameters still follow the guideline for lotus effect developed in Sect. 5.2.4. As a result, the modified biaxial structure is featured with lotus effect and a patterned upper layer in "S" shape, which serves as a track for droplets. When the surface is tilted to roll-off angle, droplets can roll off the surface following the "S" track. This controlled motion of droplets is expected to have potential applications in microfluidic system, inkjet printing, and water harvesting.

(5) **Controlling the hydrophobicity of the biaxial structure by stretching**. As shown in Fig. 5.4a, the hydrophobicity of the biaxial structure increases with the upper-layer inter-fiber distance (l) when l is less than 35 μm. This relationship can be adopted to control hydrophobicity as follows. In electrospinning, a flexible frame is attached to the collector such that the fibers are deposited onto the frame; a biaxial structure is produced with a low upper-layer l. As a result, the obtained biaxial structure is supported by the flexible frame and has a relatively low hydrophobicity. By stretching the frame along the alignment of the lower-layer fibers, the upper-layer l increases,

leading to increment of hydrophobicity, even to lotus superhydrophobicity. When the stretching is relaxed, the biaxial structure returns to its low hydrophobicity. Thus, the biaxial structure achieves a reversible transition between droplets-pinning effect to lotus effect by stretching the frame of the structure.

Furthermore, it is worth studying the change of wetting behaviors for droplets that have already been on the surface in stretching the frame. If the on-surface droplets also become easier to roll off, the biaxial structure is able to control the movement of droplets. This ability can be useful in water collection, i.e., collecting droplets by relaxing frame and transporting droplets by stretching frame.

(6) **Exploring the use of fiber-surface pores**. First, the fiber-surface pores can be filled with therapeutic compounds and then applied in wound dressing or tissue engineering. For example, a biaxial structure, where the lower-layer porous fibers are filled with drugs, can be used in wound dressing. After patching the biaxial structure on the wound, the low layer can release the filled drug to the wound. At the same time, the biaxial structure provides a self-cleaning, breathable and waterproof surface to the wound. Furthermore, porous cellulose acetate fibers can be added below the biaxial structure for the biocompatibility of cellulose acetate. Moreover, the nanoscale surface roughness on porous cellulose acetate fibers can improve cell attachment [9]. Second, the pores can be filled with phase change material for temperature control of WBS membrane based on the developed biaxial structure.

References

1. J. Xue, T. Wu, Y. Dai, Y. Xia, Electrospinning and electrospun nanofibers: methods, materials, and applications. Chem. Rev. **119**(8), 5298–5415 (2019)
2. H. Yuan, Q. Zhou, Y. Zhang. Improving fiber alignment during electrospinning, in *Electrospun Nanofibers* (2017), pp. 125–147
3. M.M. Nazemi, A. Khodabandeh, A. Hadjizadeh, Near-field electrospinning: crucial parameters, challenges, and applications. ACS Appl. Bio Mater. **5**(2), 394–412 (2022)
4. H. Aziz, M. Amrei, A. Dotivala, C. Tang, H.V. Tafreshi, Modeling Cassie droplets on superhydrophobic coatings with orthogonal fibrous structures. Colloids Surf. A **512**, 61–70 (2017)
5. X.-L. Meng, L.-S. Wan, Z.-K. Xu, Insights into the static and advancing water contact angles on surfaces anisotropised with aligned fibers: experiments and modeling. Colloids Surf. A **389**(1–3), 213–221 (2011)
6. Y. Han, Y. Xu, S. Zhang, T. Li, S. Ramakrishna, Y. Liu, Progress of improving mechanical strength of electrospun nanofibrous membranes. Macromol. Mater. Eng. **305**(11), 2000230 (2020)
7. G.-D. Kang, Y.-M. Cao, Application and modification of poly (vinylidene fluoride) (PVDF) membranes—A review. J. Membr. Sci. **463**, 145–165 (2014)
8. C. Yang, Q. Zeng, J. Huang, Z. Guo, Droplet manipulation on superhydrophobic surfaces based on external stimulation: a review. Adv. Colloid Interface Sci. 102724 (2022)

9. L. Ghasemi-Mobarakeh, M.P. Prabhakaran, M. Morshed, M.-H. Nasr-Esfahani, S. Ramakrishna, Electrospun poly (ε-caprolactone)/gelatin nanofibrous scaffolds for nerve tissue engineering. Biomaterials **29**(34), 4532–4539 (2008)

GPSR Compliance

The European Union's (EU) General Product Safety Regulation (GPSR) is a set of rules that requires consumer products to be safe and our obligations to ensure this.

If you have any concerns about our products, you can contact us on ProductSafety@springernature.com

In case Publisher is established outside the EU, the EU authorized representative is:

Springer Nature Customer Service Center GmbH
Europaplatz 3
69115 Heidelberg, Germany

The manufacturer's authorised representative in the EU is Springer
Nature Customer Service Centre GmbH, Europaplatz 3, 69115 Heidelberg,
Germany. If you have any concerns regarding our products, please
contact ProductSafety@springernature.com

Printed and bound by CPI Group (UK) Ltd, Croydon, CR0 4YY
01/10/2026
02225883-0020